Soviet market economy

DATE DUE

OCT 1 5 1992			
MAY 6 1993			
OCT 2 8 1994			
JAN 3 1995			
FEB 2 1995			
SEP 1 3 1997			
JUN 0 4 2002			

SOVIET MARKET ECONOMY

Soviet Market Economy: Challenges and Reality

edited by

Boris Z. Milner

Institute of Economics
Academy of Sciences of the USSR
Moscow, USSR

Dmitry S. Lvov

Central Economic and Mathematical Institute
Academy of Sciences of the USSR
Moscow, USSR

1991

North-Holland
Amsterdam · London · New York · Tokyo

ELSEVIER SCIENCE PUBLISHERS B.V.
Sara Burgerhartstraat 25
P.O. Box 211, 1000 AE Amsterdam, The Netherlands

Distributors for the United States and Canada:
ELSEVIER SCIENCE PUBLISHING COMPANY, INC.
655 Avenue of the Americas
New York, N.Y. 10010, U.S.A.

Library of Congress Cataloging-in-Publication Data

Soviet market economy : challenges and reality / edited by Boris Z.
Milner, Dmitry S. Lvov.
      p.   cm.
   ISBN 0-444-88979-5
   1. Soviet Union--Economic conditions--1985- 2. Soviet Union-
-Economic policy--1986- 3. Mixed economy--Soviet Union.
I. Milner, B. Z. (Boris Zakharovich) II. Lvov, Dmitrii Semenovich.
HC336.26.S684 1991
338.947--dc20   91-32621
                                                                  CIP

ISBN: 0 444 88979 5

Printed in The Netherlands

Contents

Introduction. The economic reform in the USSR: The current stage, by
B.Z. Milner   1

Economic Interrelations Between the Republics: New Approaches   3
The Logic and the Stages of the Improvement, and the Transition to the Market Economy   4
The Programme of Stabilizing the Economy: The First Step to the Market   8
The Sequence and the Stages for the Changes   10
The Social Policy During the Transition to the Market   13
The Structural and Investment Policy During the Transition to the Market   15

Chapter I. Radical economic reform: first-priority and long-term measures programme, by L.I. Abalkin   17

1. The Ultimate Goal   17
2. The Path Covered   21
3. Transition Period Options   23
4. Major Elements of the Economic Mechanism of the Transition Period   26
   4.1. Development of Comprehensive Forms of Socialist Property Ownership   26
   4.2. Restructuring of Financial-Crediting System, Financial Recovery and Money Supply
        Stabilization   27
   4.3. Wage Reform, Labor Relations and Social Support of the Population   29
   4.4. Forming a Market, Restructuring of Planning and Price-Setting   30
   4.5. Organization Structures: Transformations   32
   4.6. Restructuring of Foreign Economic Activities   34
5. Stages of the Reform   34

Chapter II. Introducing a market economy in the USSR: pressing problems, by N.Y. Petrakov   39

Chapter III. The economic system of the USSR, by S.S. Shatalin   51

1. Laws Governing the Economy of the USSR   51
2. Radical Improvement of Centralized Economic Management   60
3. The Problems of a Switchover to Real Cost-Accounting   72
4. Some Socio-Political Aspects of Radical Economic Reform   80
5. Outlook for the Development of Diverse Forms of Ownership in the USSR   86
6. Socialist-Based Motivational Mechanism for the Effective Use of Resources 96

Chapter IV. The economy at a crossroads, by D.S. Lvov and S.Yu. Glaziev   101
1. The Economic Reform and Its Tendencies   101
2. The Mechanism of Development and the Technological Structure of the Soviet Economy   112
3. How to Enhance the Economic Reform   124

Chapter V. Problems of transition to new forms of management, by
            B.Z. Milner   143

Chapter VI. The economy of a region: planning, management, cost-
            accounting, by B.Z. Milner   157
1. Basics of Regional Economics   157
2. Restructuring the Management of the Economy and the Social Development of the
   Regions   160
3. The Fundamentals and Methods of Regional Cost-Accounting   164

Chapter VII. Socialist ownership: from uniformity to multiformity, by
            L.V. Nikiforov   171
1. Moving from Nationalization to a Civic Society   171
   1.1. After effects of rejecting alternative ways of development   171
   1.2. Fundamental diversity of social and economic development   173
   1.3. The administrative-centralized system and socialism   176
   1.4. Socialism and a civic society   179
   1.5. Cooperative socialism and mixed economy systems   185
2. State Ownership: The Essence of Changes   187
   2.1. Joint-stock ownership   187
   2.2. Lease-holding relations: problems at the formative stage   189
3. Cooperation: Tactical Setbacks and Strategic Prospects   197

Chapter VIII. The current stage of reform of foreign economic relations,
            by I.P. Faminsky   203
1. The Directions of Reform and the Difficulties to be Met   203
2. The Increasing Role of Business in Foreign Economic Relations   208
3. A New Role for Trade Intermediaries   211
4. Conditions for Foreign Capital Investments   214

Chapter IX. Problems of radical reform of pricing system in the USSR, by
            Yu.V. Borozdin   217

Chapter X. Attempts to adapt the managerial bureaucracy of the USSR
            to a market economy, by I.S. Oleinik   247
1. The Nature of the USSR Bureaucracy and the Processes of its Development   248
2. Economic Consequences of the Inefficiency of the Industrial Bureaucracy   253

3. Resistance of the Bureaucratic Machine to Radical Reform   258

Chapter XI. Intensification of the Soviet economy, by V.L. Perlamutrov   263
1. The Essence of the Problem   263
2. Who is the Owner?   269
3. Democratic Choice of the Paths of Social Development   272
4. Workers as Co-owners of Enterprises   283
5. Workers as Co-owners of the Country   288
6. Priorities   294
7. Conclusion   298

Chapter XII. Efficient employment and the labor market in the USSR, by I.S. Maslova   303
1. New Developments and New Problems   303
2. Building the Labor Market and Increasing Labor Mobility   306
3. Building of the Regulated Labor Market   309
4. Greater Social Securities Against Unemployment   313

Introduction
The economic reform in the USSR: The current stage

B.Z. Milner

Professor of Economics, Acting Director of the Institute of Economics, USSR Academy of Sciences

A wide range of measures have been taken in the USSR during the past five years to promote the implementation of radical economic reforms and to improve the functioning of the Soviet economy. At the same time, the grave heritage of the administrative-and-command system, the inconsistent and the still to be decided character of the measures aimed at reforming the economy, the mistakes in managing the economy, the non-observance of laws – all these have given rise to a deep economic crisis in the country. The economy is now in an extremely dangerous zone – the old administrative system of management has been demolished, but new incentives for work under market conditions have not yet been created. Drastic measures based on public agreement are needed to stabilize the situation and accelerate the advance towards a market economy. The success or failure of the historic transformations under way in the USSR will depend directly on the progress made during the period of transition from bureaucratic socialism to a market system of economic management.

The attempts made to reform the administrative–bureaucratic system of centralized economic management made in the 1960s, 1970s and 1980s graphically demonstrated that there was no alternative to the transition to the market. Worldwide experience has proved the viability and effectiveness of the market type of economy. The interests of mankind entirely dictate a transition to such a system with the aim of developing a socially-oriented economy, bringing all production into line with consumer demand, eliminating commodity shortages and the disgrace of queues, ensuring in practice the economic freedom of citizens, setting up conditions for encouraging diligence, creativity and initiative, and high productivity.

The transition to an economic system based on market principles will help integrate our economy into the world economy, and thus allow our citizens access to all the achievements of the most advanced economies.

A difficult and radical change, which has to be effected to ensure the future of the country, will mean that the state control, dependence of the citizen on the state, levelling down, apathy and mismanagement inherent in the administrative-and-command system will be replaced by the free economic activity and the responsibility of labor collectives and each citizen for their own well-being, intensive and well-organized work and remuneration according to results.

A single economic sphere integrating all the republics and regions of the country will be formed through the All-Union market. In addition, the transition to the market will establish an economic basis for a voluntary unification of sovereign republics within the framework of a renovated and strong Union.

To make the market economy function effectively it will be necessary to create the following basic conditions during the transitional period:
– Maximum freedom for economic activity. Free commodity producers increasing their profits and consequently the national wealth will form the foundation of the economy. Acknowledgment of the social importance and all possible encouragement to the most active, skilled and talented people – workers, peasants, engineers, businessmen and production executives.
– Total responsibility of economic organization, entrepreneurs, and all workers for the results of their economic activity, based on the equality of all forms of ownership. The whole point in reforming the ownership relations is to define clearly who bears the responsibility for the outcome of economic management, to find in the course of economic development the sphere of the most effective application for each form of ownership.
– Competition of producers as a most important factor of stimulating economic activity, increasing the diversity and improving the quality of produce in conformity with the consumer demand, reducing the costs and stabilizing the prices. The development of sound competition requires the removal of monopolies from the economy parallel with the shaping of an appropriate structure of production.
– Free price formation. Market mechanisms can function effectively only on condition that most prices in the market are set freely, balancing the demand and the supply. State control over prices is permissible only in a limited sphere.
– The state's withdrawal from direct participation in economic activity (with the exception of certain specific sectors).
– Extension of market relations to those spheres where they show a higher effectiveness as compared with administrative forms of regulation. Meanwhile, a wide non-market sector will remain in the economy, comprising those activities that cannot be subjected exclusively to commercial criteria (national defense, public health, education, science, culture).
– The open character of the economy, and its consistent integration into the system of world economic trade and cooperation. All economic organizations have the right to operate in foreign economies. Foreign companies will operate in the domestic market on equal terms with all producers in compliance with the established legislation and the generally accepted international norms.

– Provision by the state, at all levels, of social guarantees to its citizens, interpreted, on the one hand, as the right to all citizens to equal opportunities so as to ensure a worthy life for themselves through work and savings, and on the other hand, as state assistance to the disabled and socially vulnerable members of society.

However, while ensuring the high economic effectiveness of production, the market requires state and public regulation, above all from the viewpoint of anticipating such negative phenomena as inflation, unemployment, excessive property differentiation, instability of production, and the uneven development of regions. By implementing a macroeconomic policy the state promotes the shaping of an environment favorable for economic activity, primarily in those directions corresponding to the public interest. The activity of state bodies in regulating the economy is based on a strict delimitation of legislative, executive and judicial powers.

A compromise programme for developing a market economy in the USSR has been worked out and will be implemented in several consecutive stages.

A prominent place in this programme is assigned to the establishment of a new constitutional system and a mechanism for the interrelations between the Union Republics whose parliaments have proclaimed their economic and political independence.

Economic Interrelations Between the Republics: New Approaches

The economic interrelations between sovereign republics are, under the new conditions, based on the recognition of state sovereignty and the equality of the republics and at the same time on the integrity of the Union as a single state, on the understanding that individual enterprises are the foundation of the economy, and the task of the state is to create the most favorable conditions for their activity.

The republics exercise the legislative regulation of the ownership, utilization and disposal of the total national wealth in their territories and it is this that forms the basis of their state sovereignty.

To realize the goals common to all the sovereign republics, joint regulations for all the republics (the Union property) have been formed under the control of Union organizations. They ensure the uniform regulation of the economic regimes on the basis of antimonopoly legislation, coordinated measures aimed against unfair competition, protection of consumer interests, and the regulation and levelling up of the starting conditions for the transition of the republics to market economies.

The republics bear the main responsibility for the development of their territories and for carrying out the economic policy, independently forming the structure of the national economic management, the system of republican and local taxes, duties and compulsory payments, and regulating the prices, the incomes and the problems pertaining to the social protection of the population.

The role of autonomous republics and other national entities will grow considerably in the economy of the USSR and the Union Republics.

In their mutual interest, and on voluntary basis, the sovereign republics enter the Union and form a single economic sphere, an All-Union market, carrying out a coordinated policy in support of free enterprise, and mutually beneficial economic ties to protect the market.

The republics jointly develop the foundations of a common economic policy and adopt legislation regulation the system of inter-republican relations, work out the procedure for resolving economic disputes and conflicts, carry out a coordinated policy with regard to prices, incomes, employment, pensions and social guarantees to citizens. They set up Union administration bodies for joint management of the spheres of activity that require a coordinated policy.

The Union budget is financed by Union (Federal) taxes; the forms of taxation and tax rates are established by agreement between the republics. The revenues for the Union budget are also formed out of other receipts from the performance by the Union of the functions within its authority. This will allow the Union to plan its expenditure, proceeding from the expected revenues, and to bear responsibility for its financial policy. Side by side with the expenditure of the departments that come within the All-Union jurisdiction, the Union budget forms the fund for regional development and support, and also makes allocations for servicing the domestic and the foreign state debt of the USSR and into the reserve fund.

To finance the foreign economic activity of the Union within the jurisdiction delegated to it, a Monetary Fund of the Union will be established, sufficient for servicing the foreign state debt of the USSR, implementing a coordinated All-Union monetary policy, providing support to directions within the All-Union jurisdiction, and forming foreign exchange reserves at the Union level.

An Inter-Republican Economic Committee will be established under the Council of Federation to coordinate the measures implemented by all the republics.

The Logic and the Stages of the Improvement, and the Transition to the Market Economy

From the very outset of the reform the accent will be made on the improvement of finance, credit and the monetary system parallel with the maintenance of the existing economic ties and material flows in the national economy. This will go hand-in-hand with the reduction of prices and inflation.

On the basis, a wide-scale transition is expected to be accomplished in market prices with due regard for the socio-economic conditions that occur.

Simultaneously, drastic measures will have to be carried out for the denationalization and demonopolization of the economy, expanding entrepreneurship and competition, so that the foundations for the engagement of the market self-regulating mechanisms are laid within a short period of time. Only these mechanisms will be able to curb the inflation and to stimulate the growth and diversification of production, the improvement of its quality and the reduction of cost. Until they start working at a sufficient strength it will be necessary to maintain a strict finance and credit policy, and in some spheres a policy of direct administrative influence to check uncontrollable inflation. This is why the structural reorganization of the economy during the transitional period will have to be confined to measures that are not overelaborate.

This difficult path must be covered in the shortest possible period of time. The experience of implementing stabilization programmes in other countries, and simulations and forecasts made for application to our conditions indicate that this period might be a year and a half or two years long. The additional authority that has been conferred upon the President of the country is intended to be valid for exactly this period.

This period will be followed by the development of sound market, an active structural and investment policy, by the growth of production and the enhancement of its effectiveness as a basis for the improvement of the life of the Soviet people.

During a comparatively brief period, in which the organs of power of the the Union, republics and regions can count on the mandate of popular trust, it will be necessary to stabilize the national economy and accomplish the transformations that will pave the way for market relations and ensure a noticeable improvement in the economy and in the life of people.

The tasks of stabilizing the economy and advancing the market should be accomplished in four stages.

During the first stage it is planned to implement a series of interrelated measures to stabilize the economy, including the improvement of finance and currency circulation through the reduction of the state budget deficit, the control of the money supply, the reorganization of the banking system and the regulation of the financial affairs of private enterprises; normalization of the consumer market; prevention of a slump in production; denationalization and privatization of property; and stabilization of foreign economic relations.

At this stage, which began in 1991, the most difficult task is to curb the inflation that is now in progress: to limit the growth of wholesale and retail prices, to effect a stage-by-stage rise of state prices of fuel, raw materials, and construction materials, and to control the level of state retail prices on mass consumption goods. All this will have be done in a manner that, on the one hand, will make it possible to exceed the budget deficit growing because of the rise in retail and wholesale prices, and, on the other, to prevent inflation from becoming uncontrollable, and at the same time to improve the conditions of the commodity–money balance for an accelerated transition to the market in the future.

The main point of the second stage is to accomplish a consistent step-by-step transition to market prices covering a wide range of technical and production means and consumer goods. Measures taken during the preceding stage must prevent their excessive growth. In future, inflationary processes will also be controlled by an austere financial and credit policy.

At this stage, retail state prices will be retained for not less than one third of all goods – fuel, raw and basic materials, which are vital for the regulation of the general level of prices and retail state prices for essential goods, which determine the minimum subsistence level of the population.

Simultaneously, denationalization will be gathering momentum, privatization of small enterprises will be under way alongside the development of the market infrastructure.

Special measures will be taken for the redistribution of capital investments and other resources in favor of industries working for the benefit of the population, thus enhancing the incentives for their expansion.

The first impact of the emergency measures and the liberalization of prices will manifest itself in an appreciable improvement of the situation in the consumer market, especially with respect to goods sold at the demand and supply prices. The tougher and the more consistent are the measures in the sphere of finance and credit policy and those aimed at the removal of peaks in demand caused by panic buying, the lower the price rise will be. Unrestricted sale of goods will be under way. It may be the first positive result of the reforms, and it must be achieved without fail.

During the transitional period, special mechanisms of social protection of the income of the population will be introduced, including its indexation depending on the dynamics of retail prices. Parallel with that, measures will be taken to give support to socially vulnerable sections of the population.

The Union Republics, the local bodies of power, may also apply different measures for the adjustment of prices, including the temporary freezing of prices for the most popular goods in the event of their excessive growth, introducing the rationing of certain goods accompanied by compensation of the losses incurred by the manufacturers and to the trade out of their budgets, and to provide social protection for the incomes of the population.

To consolidate production and the economic ties, a state system of contracts will be established. It will be responsible for placing state orders on a contractual basis, for the distribution of produce delivered on these orders and for the adjustment of its prices.

The main goal of the third stage is to stabilize, in the main, the market for both consumer goods and the means of production, expanding the sphere of market relations and developing a new system of economic ties.

Contradictory processes will be underway in the economy at this time. On the one hand, the market is expected to be increasingly flooded with goods. The market infrastructure will be developing rapidly, and the impact of entrepreneurship on the economic activity will become stronger. The utilization of material resources must become more economical, and production stocks will grow smaller. The resources released will augment the means of production for the market and further its stabilization.

The system of labor remuneration will also have to be reorganized. This will mean the abolition of regulations and the establishment of a state minimum of wages and salaries for enterprises of all ownership types. The state minimum will be based on the computation of a minimum consumer budget with due account of the new level of expenses on housing and on the whole extended range of values and services obtained at market prices at the expense of personal incomes. The real minimum of wages and salaries will become an important means of social protection of working people. Simultaneously, restrictions in the opportunities to earn money will be lifted.

Successful implementation of measures at the preceding stages will create practically all the principal conditions for a stable functioning of the economy: balanced prices and the budget, a market for consumer goods without shortages, services, a modern banking system, and a currency market.

The conditions taking shape following the expansion of the housing reform and the privatization of dwellings will be conducive to the formation of the labor market. Trade unions will consolidate their positions as defenders of working people's interests, and unions of businessmen and executives will be established. Agreements between them, with the state playing the role of regulator, will provide an opportunity to build up a labor market and to establish control over the dynamics of incomes and prices.

Finance and credit restrictions will be removed parallel with the development of competition, entrepreneurship and the stabilization of prices. Taxes and interest rates on loans will be subject to changes to enhance business activity and encourage investment.

The main goals of the fourth stage are to consolidate the stabilization of the economy and finances, to improve the consumer market, and what is most important – to speedily mold the competitive market environment necessary for effective functioning of the self-regulatory mechanisms peculiar to the market.

Substantial advancement in the demonopolization of the economy, in denationalization and privatization is a necessity at this stage.

Prerequisites must be created for enhancing the economic activity, above all in the food industry and in light industries, in the agrarian sector and in the service sphere.

The prevalence of prices reflecting demand and supply in combination with a balanced budget will establish preconditions for the solution of the key problem of transition to a market economy – the problem of the rouble's domestic convertibility. The main point of this is to afford all domestic enterprises and foreign companies operating in the USSR an opportunity to sell and buy freely at a market rate the currency needed for current economic operations.

The domestic convertibility of the rouble will create ample opportunities for the inflow of foreign investments vital for the structural reorganization in the country and also for the expansion of competition on the domestic market and for the eradication of monopolies. For the conditions of our country this is an important prerequisite for the market mechanism to start working at full strength. Gathering momentum is a structural reorganization of the economy that will for the first time be based on the market mechanism and on attracting foreign investments, technologies and management experience.

The Programme of Stabilizing the Economy: The First Step to the Market

Improvement of the financial system and the monetary circulation is the key target of stabilization. The main point is to eliminate the surplus money in circulation, to bring it in line with the commodity resources, and to consolidate the rouble. At the same time the USSR State Bank stops financing the budget deficits, and its expenditure is sharply reduced. All projects far from completion are suspended, except for those intended for the expansion of consumer goods production, housing construction and development in the social sphere. Expenditures for the needs of the USSR Defense Ministry and the State Security Committee are curtailed, particularly through the reduction of arms purchases and military construction. Payments of subsidies to enterprises are reduced. Expenses on the maintenance of the state apparatus are brought down to a minimum. The system of taxation is changed.

Of decisive significance for the improvement of the monetary circulation, the prevention of inflation, and the establishment of tougher financial restrictions is the transformation of the banking system and the development of effective policies for controlling the money supply.

With this aim in view, specialized state banks are transformed into joint-stock commercial banks operating on a par with other commercial banks. They are responsible for the direct credit and accounting service of the national economy.

The USSR state banking system is transformed into a Reserve System of the Union consisting of the USSR State Bank and the central banks of all the Union Republics. One set of coordinated and universally binding rules is established for regulating credit and money supply, including reserve requirements for commercial banks, interest rates, etc.

The interest rates rise. In future the level of interest rates will be determined by the supply and demand on the credit market and by the policy of the Reserve System. Thus, the currency becomes expensive and acquires the qualities that are indispensable for it to circulate in a market economy.

Many enterprises will experience financial difficulties as a result of these measures. To facilitate their adaptation to new conditions, give them a chance to accumulate their own current assets and raise their effectiveness, it is planned to:

– immediately denationalize, and privatize such enterprises and break them into smaller units;

– set up stabilization funds;

– develop commercial credit between enterprises as an element of a new wholesale market with a view to reducing the reliance upon bank credit;

– stimulate issuing of bonds for augmenting current assets;

– close hopelessly ineffective enterprises.

The transition to the market economy and the effective functioning of the self-adjusting market mechanism are possible only on condition that the bulk of manufacturers – enterprises, organizations and citizens – enjoy the freedom of economic activity and entrepreneurship. They must have a free hand to utilize their leased or owned property, independently define a production programme, choose suppliers and consumers, set the prices, dispose of the profit after the taxes, and handle other issues related to economic activity and the development of production.

This will create the necessary conditions for the functioning of the market mechanism and mobilize the human potential – a major resource capable in our conditions of producing, within a comparatively brief period, an effect based on people's aspiration to obtain normal living standards through their own work and skills.

The molding of free commodity producers as a major element of the market economy requires a speedy withdrawal of most enterprises from state control, and their privatization. Privatization is not necessarily interpreted as a transition exclusively to private ownership, but rather as a more general process of the change of proprietorship through a transfer or sale of state property on different terms to collectives, cooperatives, share holders, foreign companies, or private individuals. These types of economic management meet the demands of the market economy best of all. They are not only independent in their activity, but are also responsible for its results in economic terms, both in their current incomes and in its capital value. This ensures the rational use of the resources and restrains their unfounded redistribution for consumption to the detriment of the modernization of the production potential.

The denationalization with the transformation of ownership rights can be accomplished by different methods. Concrete trends in the denationalization will be determined depending on the peculiarities of republics and regions, on the specific features of an economic sector or industry, on the size of an enterprise, on the state of funds and on other factors.

Immediate measures aimed at denationalization and privatization are carried out in the spheres where the functioning of non-state structures is most expedient, that is in trade, public catering, consumer services, repair and construction organizations and small enterprises in other sectors. At the same time, already in 1991 large and medium-size enterprises in various industries and other spheres will begin to be transformed into joint-stock companies and associations.

The present book value of the country's capital funds (minus the value of land, mineral wealth, forests and the private property of citizens) amounts to about three trillion roubles and, with allowances for wear and tear, to about two trillion roubles. The current assets of enterprises and associations total 800 billion roubles. Nearly 90 percent of property is owned by the state.

This is why the process of denationalization may stretch over a rather long period. Various methods will be probed to accelerate this process in the interests of all citizens.

Special measures will be taken to demonopolize the national economy and encourage competition. Their main aim is to prevent, limit and check the use by the participants of the economic circulation of their dominant position on the market and to bar unfair competition.

This work will be headed by a specially established Antimonopoly Committee. Similar bodies are set up in the Union Republics.

The Sequence and the Stages for the Changes

The agrarian reform occupies a special place in the programme of developing the market economy. In the course of its implementation conditions will be created for the coexistence of various forms of property, including private property, and a multistructured economy will be formed in the agrarian sector. Simultaneously, administrative structures of state control over agricultural production will be eliminated.

The supreme bodies of republican state power will set up committees on the agrarian reform responsible for making inventory and evaluating arable lands, exposing ineffectively used lands, performing land tenure, consultative and controlling functions.

Ineffective collective and state farms are in the process of transformation, with their lands being fully or partially transferred to cooperatives, leaseholders, peasant farms, industrial and other enterprises to organize commodity agricultural production, and also to set up individual small holdings, gardening and kitchen gardening.

Republics launch programmes in support of commodity (farm) production and other new forms of economic activity in the countryside, and the creation for them of an economic and legal framework comparable with other agro-industrial organizations. Particular attention is devoted to their development in regions where labor is in short supply.

Measures for demonopolizing the sphere of purchasing and processing of agricultural produce are taken in the Union and autonomous republics. Small enterprises, equipped with up-to-date machinery, will be directly established on a priority basis in the agricultural areas. Processing plants will continuously expand their operations as free commodity producers, purchasing raw materials and selling their produce on market terms. The bodies of power and management give the utmost encouragement to the establishment of cooperative wholesale and intermediary services and firms, purchasing and sale associations, and sale and purchasing cooperatives. Large processing enterprises will be consistently transformed into joint-stock societies.

The price formation system is another important element of the market reform. Free market prices are an integral element of a market economy. To advance to them as smoothly as possible, in the first place regarding the prices for consumer goods and services, it is planned to free them stage by stage from administrative control. During the first stage control will be lifted from the prices on the group of nonessential goods.

At the same time, to ensure social protection of the population, provisions are made for maintaining state prices on goods and services that make up the basis for the subsistence wages of families (the set of food and industrial products for these purposes will be determined). Local price rises for certain goods may be affected by the decisions of individual republics, allowing for local conditions. In such cases, republican and local bodies set up compensation schemes to protect the incomes of the population according to their potential, place orders for additional production of these goods, and introduce their rationing when necessary.

The lifting of control over prices must and will be accompanied by a tough financial and credit policy, accelerated rates of denationalization and demonopolization of the economy, development of competition and an entrepreneurial market infrastructure.

In the event of an excessive rise of prices reieased from control earlier, republican and local authorities may temporarily set limits on them again.

During the transitional period, the state cannot immediately renounce the policy of support to low retail prices on certain goods and, consequently, stop subsidizing them. However, it is necessary to change the procedure of subsidizing the prices of goods radically, so that it will be profitable to produce and sell them and, consequently, cheep goods will not suddenly become in short supply. Under the existing considerable financial and monetary imbalances, the transition to contract prices can give rise to excessive inflation, overestimation of costs (so as to cut down the income deductions in the budget), and lack of interest in the growth of production due to low profitability.

The following measures are planned to be carried out to prevent these negative trends:
– to take urgent measures to regulate the finance and credit system with a view to drastically limiting the monetary weight at enterprises and in organizations. Strict financial restrictions for buyers combined with the difficulties experienced by suppliers in the sale of produce will make contract prices fall steeply, and check inflation;
– to set as soon as possible a sounder level of state prices for fuel and power resources, the most important raw materials, construction materials, and other products of production and technical value, on which fixed prices are to be retained.

The transition to the market requires the rise and development in the USSR of an open type of the economy, interrelating and competing with the world economy. Therefore, the implemented reforms envisage preservation of the indivisibility of the customs territory, the monetary system, the foundations of the investment regime and the foreign economic policy of the country parallel with the extension of powers of the Union Republics in these spheres. For these purposes a national body will be established to control foreign economic relations, with the Union Republics being directly represented on its board.

Individual enterprises becomes the key figures in foreign economic activity during the transition to the market. It is planned to denationalize and decentralize the foreign economic activity and pass commercial operations on to the level of enterprises, ensuring the latter's business independence and centering the functions of state organs on regulation and stimulation of these operations.

Oil, natural gas, gold, diamonds and precious stones, special equipment, and, possibly, some other goods on a list coordinated with the Union Republics compose an All-Union export resource and are sold separately from everything else.

Departmental foreign economic associations are transformed into mediator companies, including joint-stock ones, and, when necessary, are broken up into smaller units and diversified.

The management of relations with foreign economies is now exercised with the aid of legal and economic levers (custom duties, taxes, exchange rates, banking interest rate). A new customs tariff of the USSR is introduced to regulate foreign competition on the domestic market. The list of centrally licenced export goods is reduced to the minimum. An expanded system is established to control the observance of the foreign economic legislation and the rules of competition; customs and auditing services are consolidated.

Purpose-oriented foreign capital is attracted into the country. Its possible forms of investment are expanded (including enterprises in its total ownership, concessions, zones of free enterprise), their priority trends are established. Foreign investments are gradually turned into the main channel for bringing in foreign capital.

Stabilization of domestic finance, new prices and the rouble's exchange rate, its domestic convertibility on the home wholesale and monetary market serves as a basis for a step-by-step transition to the convertibility of the Soviet currency into foreign currencies. Having started with the operations of foreign capital, it afterwards controls all the current operations of the country's balance of payments.

The Social Policy During the Transition to the Market

A socially oriented market economy is an economic system where each group and social section of the population enjoys broad opportunities for the realization of its vital capacities and demands on the basis of free labor and growing personal incomes.

An opportunity to have a free choice of where and in which forms of economic activity one can apply his knowledge and experience is a major factor in strengthening the social vitality of those involved in the national economy.

The new social policy will ensure the growth of personal earned incomes, the increase or their role in satisfying the social and everyday needs of the population and the elimination on this basis of a dependence culture. A guaranteed level of social values, which should be regarded as a minimum one, will be provided for all sections of the population through social consumption funds. And a higher level of consumption, usually on a free basis, will be provided to economically unproductive citizens – children, invalids, pensioners. All able-bodied persons must raise their living standards above all by their own labor and personal income.

The consistent growth of labor costs will be an important consequence of the new social policy. Wages will more and more correctly reflect the real costs of reproduction of skilled labor power, thus making investments in education, training, cultural and social development of the personality most effective.

In the process of reforming labor remuneration, state tariffs are regarded as the minimum guaranteed remuneration for the labor of workers with the corresponding qualification level. From this standpoint they are binding for application all over the country and for all enterprises, irrespective of the form of business. Republics and enterprises may use their own discretion in setting labor rates at the expense of their own resources, but not lower than set by the state.

In this sense the state system of tariffs will become a factor of social protection for working people and a powerful incentive for raising the effectiveness of production.

The new tariff system must cover both productive and non-productive sectors of the national economy, so that cultural, public-health and education workers will have the same guarantees in labor remuneration and the same level thereof as the workers of the productive sphere.

At the same time, a new system of hiring the executives of enterprises under a contract is introduced, where the state tariff also acts as a minimum of labor remuneration.

Enterprises settle all other issues related to labor remuneration – forms and systems of payment, bonuses, rewards, increments, etc. – independently, without interference on the part of state organs. Individual earnings are not limited, being regulated only by the income tax.

Dividends on the shares of enterprises and other returns on property become a new source of income. By purchasing shares, working people turn into masters of their plants, factories, state farms and other enterprises, thus getting additional incentives for highly productive labor.

The higher effectiveness of production, the liquidation of hopeless and not sufficiently profitable enterprises, the establishment of new industries and sectors will result in a redistribution of the work force, in the first place into the service sphere, where new jobs will be created and into cooperatives and the private sector. A certain number of workers will stay temporarily outside social production. The task is to reduce to a minimum a person's period of unemployment.

Special state employment services will be established within the local Soviet republics, which will not only seek out jobs, but also further the creation of new jobs, training and retraining, carry on vocational guidance and, finally, they will assume responsibility for the material security of the temporarily unemployed. It is a matter of principle that the employment service will render its assistance free of charge.

Ad hoc programmes of employment will be implemented at an All-Union and at a republican level, in the first place in labor-excessive regions. A fund for aiding the employment of the population will be established.

The current system of material security of those who became unemployment is planned to be supplemented with direct unemployment allowances, when for some reason or other a worker cannot be quickly provided with a job or retrained.

The Structural and Investment Policy During the Transition to the Market

The formation of a modern economic structure is essential during the transition to a market economy and a precondition for its successful functioning. The situation is really dramatic, because the scale of the changes means that any attempts to switch over to an economy based on market mechanisms without a simultaneous restructuring is virtually impossible.

The key directions of the structural reorganization are as follows:
– the formation of a developed consumption sector of the economy as a precondition for mobilizing the social reserves of the economic growth;
– the elimination of the structural and technological imbalances in the national economy, which are the main sources for the perpetuation of an economy that wastes resources. The structural and investment policy must ensure the reduction of ineffective allocation of resources with a simultaneous concentration of effort on the spheres that bring maximum returns and are fully consistent with the task of the social reorientation of the economy.

The economic initiative of enterprises and the entrepreneurial spirit or the population, which are in need of effective legal and economic guarantees, must become the driving force of the structural reorganization in the economy. Supported by state regulation measures, this powerful factor should be in the first place directed at the repletion of the consumer market.

The transition to market relations and the latter's effective functioning are feasible only under a strong and well organized state power. This is proved by the experience of all countries that have succeeded in building developed market economies and attaining high living standards for the population.

At the very outset of the transition to the market it is planned to switch over to new structures of the state and economic administration oriented to the formation of a market economy and ensuring an effective interaction of All-Union, republican and local bodies. The elaboration and implementation of economic reforms and the programmes of transition to market relations must immediately be turned into a particular function of state administration.

In the first place it is planned to make full use of the USSR President's authorities that were recently conferred on him by the Supreme Soviet of the USSR. In this situation it would be desirable to enhance the role of the Council of the Federation, which must work out and implement, via All-Union and republican authorities, the decisions coordinated among the republics. For this purpose, an Inter-Republican Economic Committee comprising the representatives of republics, specialists and scientists will be formed under the Council of the Federation.

The new duties and higher responsibilities of the President and the Council of Federation require further consideration of the structure and functions of the executive power, including its highest echelons. The structure of the organs of state power and administration at the All-Union level will be determined before the Union Treaty is signed.

A number of All-Union and Republican bodies entrusted with new functions engendered by the transition to the market will be set up in the near future.

A state contractual system will have to be developed to include a complex of organization responsible for placing orders for products required by the state, selecting contractors, and working out and signing contracts. It is also necessary to establish a Fund of the State Property of the USSR, a State Inspectorate for Controlling Securities, a State Inspectorate on Prices and Standards, an Investment Fund for Regional Development, an All-Union Monetary Fund, an Economic Stabilization Fund, a Fund in Support of Employment of the Population, an Anti-Monopoly Committee of the USSR, a State Insurance Supervisory body of the USSR, a Pension Fund of the USSR, a Committee for Small Enterprises and Entrepreneurial activities, as well as the corresponding organizations in the Union Republics.

The republics will have to make an inventory and expert evaluation of state property, arrange auctions for selling this to citizens with the right of ownership; in the first place it concerns the objects of unfinished construction, unmounted equipment, construction materials, transportation means, enterprises engaged in trade, public catering, and consumer services.

A specific task of the republican authorities is to work out the main principles and start implementing agrarian and housing reforms, to put into practice the programme in support of entrepreneurial activities, the agro-industrial complex, and the non-market sector of the economy.

The programme for a transition to the market has already been drawn up, but all measures aimed at the improvement of the economy may fail to produce the expected effect, unless political stability is maintained in society, along with the willing and effective cooperation of all political parties, organs of power and the administration. Without such a consensus and support on the part of the population, any programme of economic reforms, no matter how good it is, may fail.

Chapter I

Radical economic reform: first-priority and long-term measures programme

L.I. Abalkin

Academician, Deputy Chairman of the Council of Ministers of the USSR

A radical economic reform is basic to restructuring (perestroika); its essence, in a way. But four years of strenuous efforts have hardly brought the success that was so keenly anticipated, and the situation gets worse and worse. People get more and more worried, their concern being a fertile soil for extremists of both left and right.

Thus, reform is required as a matter of great urgency, and consideration must be given to its objectives and goals, and the course of its implementation. It is essential that positive and realistic steps are taken in order to overcome the crisis in the economy of the USSR.

1. The Ultimate Goal

The ultimate goal of radical economic reform is a sound economic system which can ensure highly efficient production, modern living standards, social justice and the means to solve the most urgent ecological problems. Only then can one speak of the completion of "perestroika" and the success of reform. Only then will prerequisites for the development of Soviet society in terms of freedom and democracy be created. Only then will the ideals and standards of socialism have been achieved.

To achieve this goal, an economic system is required which: (1) provides real incentives for labor and enterprise combined with a high degree of organization and discipline; and (2) coordinates the activities of all participants in the production process, so that energy, efforts and materials will not be wasted. The aim is that everyone produces what society needs, using the resources which give best results and most advantageous outcomes. The concept of socialism as Vladimir I. Lenin saw it in the last years of his life, with support from the analysis of our revolution and the new economic policy (NEP), the experience of socialist and other countries, and the experience of our restructuring and long years of in-depth studies, permits an outline of the principal features of a new economic system within the framework of socialist options, which are as follows:

(1) Establishment of diversity of the ways in which property can be held, these ways to be equally legitimate, together with competition as the basic philosophy and guarantee of economic freedom of citizens, who can then make the best use of their own abilities, under powerful individual and collective economic motivations.

Diversity of ways of holding property is in no way a transitional, but a normal state of economy. It opens the possibility of doing away with alienation of the working people from the means of production, from power and participation in managing economic affairs. The initiative from the people, for systematic de-nationalization of property, in favor of developing leasing, cooperation, peasant and farmer's property, public property, share-holding societies and other property owning entities, appears entirely natural. These new formations enable persistent and logical implementation of principles of complete cost-accounting, economic reliability and responsibility, self-financing and self-management and each ought to be employed where it best suits the possible degree of socializing of the production process and where it best promotes highest possible efficiency.

Only a radical change in property-holding patterns can lead to the solution of the key problem, which has not yet been found in the socialist economy: how to elaborate, in the long term, enterprises and work teams which will collectively determine optimal balances between consuming and saving, between product innovation and the expansion of production.

This will supersede the dogma of incompatibility of socialism and the generation of income from property. Historic experience has shown that property-generated income should be treated as money, having with profits and other market phenomena a common economic content and being able to promote various socially useful ways of production. Without them it is difficult to create incentives for saving, and for increasing efficiency through capital investment, which both ensure a balanced consumers' market.

At the same time, property-derived incomes and profits, in their socialized forms, including interests from investments and bonds, dividends, interest on share and charter capital, insurance premiums and pension and investment contributions, can be efficiently regulated centrally, which will help avoid the socially unacceptable growth of unearned incomes and profits, and the possibility of exploitation op people by each other.

(2) Stemming from the labor principle of property, production of goods, its production and distribution can be related by the formula: to earn and to share. The chief principle for measuring consumption rights is the labor contribution, the participation in creating the product. However, currently the principle of the administrative–authoritarian pattern of management is to take away from

manufacturers their product in order to distribute it. This gives the latter an exclusive position when it is advantageous to give less and to get more, i.e. they are paid not what they earn, but what they can squeeze, out of the community's resources. Priority of labor is what is preached, priority of distribution is what is practiced. This is the gist of the economic thinking of a dependent.

The new economic system should re-install the natural way of things, when the economy would stand on its own feet, head up, when everyone involved would discover and understand that one can consume only what one has earned, one can sell what one has produced.

(3) The market is to be the chief method for coordinating the activities of the participants in social production. From our own experience we could see that there is no alternative to the market mechanism in this role. This is also the most democratic way to regulate economic activity. The attempt to create an economic system based on the coexistence, in opposition, of a plan and a market resulted in an authoritarian management system giving rise to bureaucracy and irresponsibility and leading to waste of material and resources.

Diversifying the ways in which property can be held leads to economic independence and economic responsibility and at the same time it sets the prerequisites of the normal functioning of a market mechanism. But in creating a market one cannot stop half-way: a market is capable of effective functioning only under conditions of free unregulated prices and economic competition. A plan-oriented regulation of the market is essential, but it should be carried out not counter to its laws, but on the basis of them.

A major aspect of the market economy is a financial market, i.e. a market for stocks and bonds, which will divert social resources quickly into the areas of most efficient application and in the long run will encourage savings, scientific and technological advance.

It is necessary to admit the existence of a labor market under socialism. In fact it has always existed and has had a major impact on the level of payments (wages and salaries). Attempts to direct labor through administrative methods to regulate incomes and establish differentials among them resulted only in limiting individual freedom, and through levelling down preventing people from selecting work where their individual abilities could best be revealed and employed.

Under socialism however, the labor market will be defined and regulated through a comprehensive system of social guarantees, backed up by state protection of the rights of the working people, as well as by representation of all the categories of membership of units of production, co-ownership rather than employment.

(4) A system of social guarantees must be developed to ensure equal opportunities for all citizens for harmonious development, with comprehensive implementation of principles of humanity and social justice.

An important feature of the new socialist economic system is a balance between economic efficiency and social justice. Socialist economics evolved from the concept that social justice, represented by the principles of socialized property, of elimination of exploitation of man by man, and of distribution in accordance with the labor invested, will ensure economic efficiency purely of itself. Experience has proved how unrealistic this presumption was. A contradiction between economic efficiency and social justice is of objective truth. Dialectically, a society can afford to be more just the more efficient its economics is. But at the same time, social justice and humanity can be a better way to achieve profit in the long run, than can the application of purely rational profit–loss accounting.

(5) A planned state regulating the economy on the basis of perspective (future-oriented) economic and social planning by:
– directly managing state (republican, national and communal (municipal)) enterprises, natural resources and property;
– placing and assigning state orders on competitive and contract basis;
– state investments and subsidies;
– price regulating;
– taxation, tax exemptions and financial measures;
– regulating the money supply and crediting investments through banking interest rates, reserve standards, currency exchange rates and other methods;
– making use of the state reserves of goods to stabilize the market and prices;
– legislative regulation of economic activities, to prevent and avoid monopolistic steps, in particular, encouraging sound competition, protection of customers' interests and environmental protection;
– the development of a social security system in parallel with the making up and implementing corresponding programmes;
– an ever-greater role in regulating economic and social processes will be reserved for non-state economic structures, democratic public institutes, such as consumers' associations, manufacturers' associations, ecological, scientific-technological and other public organizations.

These are, briefly, the principal features of the new economic system. Time and experience will alter and enrich our ideas on it. So it will change with time.

But right now one can say that the system, though in full accord with the principles of socialism, radically differs from the administrative–authoritarian type of management. Improvement of the present mechanism of economy management, or substituting some obsolete parts of it by dismantling the faulty unit and replacing it with a newer modification of the same, are not being discussed. Rather, an existing system with a defective element ought to be completely ousted and replaced by another one, equally self-contained and self-consistent, and incompatible with the former due to their different philosophies. Hence, the radical character of the economic reform and the inevitable difficulties connected with introducing it and carrying it through.

2. The Path Covered

The past four years have been exceptionally eventful from the point of view of publishing and explaining the basic rationale of the economic reforms launched by the Party aimed at renovating society, and creating awareness of their profound character, and exceptionally complex content. This accounts largely for the inconsistency of decisions adopted, and explains why some aspects of the economics, such as finance and monetary circulation, as well as acceptance of the power and prestige of authoritarian ways of management have been overlooked.

A critical estimation of the distance covered makes it possible to distinguish several stages in this seemingly homogeneous phenomenon, permitting understanding of the necessity of complete demolition of the authoritarian-management style system.

In 1985–1986 the principal stress was laid on scientific–technological advance. It was believed that it is possible to overpower and overcome negative tendencies and achieve the essential acceleration of development, preserving the old economic system in its entirety. This advance, supported by the activated energies of the people (behavioral aspect) was believed to be sufficient to ensure fast growth of production of quality goods to world standards, a penetration of the world market, and as a result a noticeable increase in the standard of living of the Soviet people.

It was also meant to put an emphasis on economic incentives. There was no end to economic experiments of all kinds, with numerous new methods of economic management being introduced. They were all, however, just like the economic transformations of the mid-sixties or the late seventies, when the greatest problem was regarded to be the necessity of establishing new estimation indices and fund-formation methods at individual enterprises. Ever more decisive were measures undertaken by the centralized management: all kinds of production associations were dismantled, the structures of ministerial level bodies were changed, the agro-industrial complex was formed, Gosstroy (State Construction Committee) was reshuffled.

At the time it was not realized that it is impossible to create effective economic incentives under the existing conditions of a cumbersome production structure featuring little, if any flexibility and an economic system rigidly tied to that structure. It is little wonder that the goal of accelerating socio-economic development of the country proved unattainable.

1987–1988 brought awareness of the necessity of more radical and more effective changes in economics. Technocratic and administrative approaches were seen as limited and one-sided. This led to adoption of a concept of radical restructuring of directions of the economy, and reassessment of roles of personnel and of management.

This was a significant step towards a complex and comprehensive restructuring of economics. The goals and steps outlined at the Plenary session of the Central Committee of the CPSU in June 1987 put a special emphasis on the enhanced rights and the increased economic responsibility and independence as a major production unit of commodities of a socialist enterprise. This new emphasis determined inevitable changes in planning, material and technical supplies, price-formation, finance and crediting mechanisms as well as in organizational management structures.

However, a distinguishing feature of the policies adopted was their indecisiveness, the fact they stopped half way, compromising between the existing, dominant opinions of the principles of the socialist system of management and the undoubtedly necessary new approaches.

This was best demonstrated in the Law on establishment of a socialist enterprise (association). Its practical implementation soon revealed its inner contradictions and the superfluous nature of many of its clauses. Besides, a new economic environment was never set up, and the insignificant minor changes in planning were not able to compensate for the failure of financial recovery and to overcome the drawbacks of the former system of pricing. The administrative methods of regulating economic activities had weakened, while the economic methods did not exist at all. This resulted in serious adverse effects on production, worsening of shortages and imbalance of profit and loss. Despite all that, there were no conclusions made about economic de-stabilization. The plan for the twelfth five-year plan period remained unchanged.

This was an attempt at radical change: the Law on Cooperation has stimulated the development of a sector of economics. But a genuine economic success was never achieved. The experience of cooperative movement development showed convincingly that under the existing system of practically complete state control of all economic institutes and of the administrative system of management, any other economic model, based on another set of economic principles would be necessarily rejected or would unwillingly become criminal. Cooperative movement became discredited in the eyes of the public.

The resulting situation was caused to a great degree by the existing stereotypes which the country failed to destroy, testifying to the power of the administrative method of management at all levels. The essential aspect of this problem was expressed in the Resolution of the 19th Party Conference which stated that radical economic reform and renovation of all aspects of social life are impossible without democratization, without changing the political system.

1989 was the year of the Congress of People's Deputies which confirmed this conclusion and served as the starting point of political reform signifying a new stage in restructuring economics.

At present we are living through a highly complicated phase of this restructuring: the unprecedentedly active politicization of democratic masses leads to actual transformations in economic and political structures. It is especially necessary in these conditions to check the operation and verify the plotting of the route.

3. Transition Period Options

The importance of the existing situation, and its critical nature in terms of possible consequences are revealed by the fact that we are faced today with an option. Various ways ahead are possible, that would tell differently on the development of the country. Each option is backed by certain social forces. The question now is as follows: to choose the way and ensure that the selected way for all its apparently progressive trends would not prove a dead end.

Generally speaking, the most complicated part of any reform is implementing the transition towards the new economic regime. For us the choice is significantly predetermined by the existing economic and socio-political conditions, by the onerous inheritance of the past. The gross distortions of the economic system consequent on the dominance of large-scale monopoly manufacturers, the chaos of financial and monetary systems, inadequate personnel training, loyalty of large groups of the working people to "level-out" distribution, all these factors have been joined by new ones: economic imbalances have been aggravated, the situation in the consumers' market has grown noticeably worse, the population's mistrust towards the state policy has increased, unreasonable demand campaigns have become more frequent. A considerable part of the working people has lost faith in the effectiveness of the reform and this has increased socio-political tension.

And the choice is to be made in so tense a situation. Whatever it is, however, we should remember one thing and very firmly, too: there are no fast and easy ways towards success. Any option adopted as a course of a further reform brings with it the necessity to carry out a number of unpopular and painful measures, each one likely to contribute to the tension in society. So it is vital to withstand the pressure of the moment, not to give in to current and fleeting feelings.

One can't help observing an alternative has been proposed to wholesale economic reform; which states there is no need for a market economy, administrative measures and sanctions ought to be made more strict, cooperatives ought to be closed, the rise of prices ought to be forbidden, etc. In the extremely dramatic situation prevailing now, this programme for all its economic inconsistency finds support among a certain part of the population inspired and agitated by the slogans of the demagogues. That is why it should be said definitely and finally: this alternative policy would throw us back considerably without solving the smallest of our problems. Economic reform has no realistic alternative. In this respect the choice has been made.

Many of the alternative ways of achieving reform include a number of measures, common to the reform at large: those leading to financial recovery, economic theory, large-scale re-structuring aimed at increased production of consumer goods and services, a cardinal shift in the assessment of property as such.

The drama of the situation calls for extraordinary measures. These could include the patching up of the budget deficit, where first steps have already been taken, a considerable increase of centralized production investment, measures to curb unjustified growth of incomes, an increase in the production of consumer goods. The scale of the efforts applied is not sufficient as yet.

When selecting additional measures, one can follow one of the two ways: emphasis may be placed either on administrative measures (with an increase of obligatory quotas to fulfil, an increased emphasis on rationed distribution of material resources and consumer goods) or on economic methods, primarily in the financial and money-credit spheres. Doubtless, in the short term, reform will require both economic and administrative limiting measures. However, well planned and thorough economic measures, painful though they may be, and causing a considerable opposition, will take us out of the crisis sooner.

Three options of the strategy for further reform have been put forward:

Option One is conservative, based on the strategy of a gradual growing-into the new system of economics. Urgent tasks would be solved by methods which minimize the negative consequences of the measures taken, thus avoiding conflicts, but such methods do not take into account long-term development and the philosophy of complete reform. Extraordinary measures in this option therefore have an administrative, forbidding character: examples include price freezes, limiting cooperatives in their activities, struggle against monopoly supplies by forbidding them to decrease volume of production and to refuse to make contracts, and the like.

The rate of financial recovery would be limited by inefficient enterprises which would thus be protected. A transition of state-owned enterprises into other forms of membership would occur slowly, and with approval of management bodies.

The satisfying of the consumer market is believed to be possible at the expense of producing goods and services at stable prices. A decrease of guaranteed state orders and of the importance of centralized distribution of products, would postpone forming of the market till the first signs of success in financial recovery and in restructuring of the production pattern. The advantage of this option lies in enterprises, population and management bodies being given an opportunity to slowly and gradually adapt themselves to the changing conditions. Here the old pattern finds ways to live side by side with the gradually introduced new forms and the counteracting forces have time to consolidate. Current and locally-oriented problems are likely to get the upper hand of the strategic purposes. Besides, the long time needed to restructure in economics implies the necessity to put up with all the discomforts of a "house under repair". In any way, a system lacking integration in its perception or presentation cannot be efficient. Tangible results of the reform would be put off till considerably later.

Delay and the spacing out of serious steps to set up a market are justified by an attempt to avoid a social tension connected with the increase of prices. However, more dangerous is the fact that the expected improvement of the situation may never come as a result of using this option, since curbing prices and incomes will further weaken demand thus promoting a decrease in production and an ever growing deficit. It is hard to expect the fulfilment of a high planned target to increase the production capacity of consumer goods since much targets do not actually take into account the real possibilities of an enterprise.

Option Two implies a radical reform which will involve very energetic suppression of all approved, traditional and official restrictive practices; one-time decisive abolition of all limits of boundaries in the market without any preliminary preparation of the system of its state regulation. The launching of this market mechanism would usually be connected with a complete or practically complete refusal to control prices and incomes accompanied by strict measures to reduce public expenditure and the money supply.

State-owned enterprises would collectively introduce leasing as a form of ownership, and get transformed into public share-holding companies, cooperatives, and other associations. The ministries would be dissolved, and state orders placed only on the contract basis to be financed from the budget.

Consequently, the task is to ensure a comprehensive set of actions to produce within the shortest possible time a new economic system and obtain the long-expected results.

However, a favorable result cannot be taken for granted. A long stagnation in the economy cannot be completely ruled out as a result of the measures undertaken and would be accompanied by socio-political shifts caused by the disadvantages of the method. The most significant unfavorable results would be prolonged galloping inflation, painful to the consumer and allowing the manufacturers to increase their income due to price growth, stock and capital appreciation without actually expanding or improving production. Bankruptcy of many enterprises, the breaking up of existing and consolidated ties are possible. A considerable decrease of production accompanied by sustained and large-scale unemployment are possible, lowering the standard of living with a strong differentiation of incomes of various layers of the population.

The Third Option is moderately radical. The essence of the method is to stimulate the forming of a market economy through a powerful combination of measures and to create simultaneously a regulatory mechanism for the process. This, certainly, will require a certain preparation but it cannot and must not take longer than a year. Then a follow-on programme of consolidating and developing a new economic system would be implemented, its major element being a step-by-step removal of control over prices further enhanced by a strongly developed social security system which, still at the preparatory stage, would ensure incomes and prices are coordinated.

A complex of simultaneous reforms would include: laws changing the relations of property, taxation, saving and lending systems, wage reforms, social security reform, a complete change of employment pattern and introduction of unemployment benefits, fixed incomes to be indexed to the cost of living, the beginning of the price reform, introduction of a system of stable, fixed, regulated and free prices, transformation of planning purchasing systems.

The state will then get an opportunity to control the process of restructuring the economy using a combined and pragmatic policy of prices and incomes, state orders and credit limits. Manipulating prices and incomes could support production incentives and avoid excessive tension in society. If these measures cause accelerated inflation rates, rigid financial measures would be introduced, credit limits and restriction of money supply imposed, this being the essence of the monitored inflation policy. This is the lesser evil, avoiding much greater damage and achieving the goal.

Thus, this option lets controlled price growth curb the increase of the deficit and slowdown in production, opening a realistic way towards formation of the market economy. A simultaneous implementation of a complete set of large-scale and thoroughly planned and prepared and coordinated measures would encourage an atmosphere of breakthrough, of a serious step forward, a turn in the course of restructuring and reform. An effective system of social security will cushion the difficulties of the transition period, and facilitate the adjustment of the working people to market economy conditions.

The options considered present an almost complete spectrum of possible strategies of implementing economic reform. The third option is the best.

4. Major Elements of the Economic Mechanism of the Transition Period

4.1. Development of Comprehensive Forms of Socialist Property Ownership

The central element of the reform to be taken up immediately is a set of practical measures to develop the diversity of the ways in which property can be held. Legislation considered at present in the Supreme Soviet sessions should provide legal and economic prerequisites aimed at an accelerated transformation of state enterprises into leased, cooperative, share-holding, public and based on an individual's activities, mixed, joint with a foreign stake in the share capital, as well as other, enterprises.

For this purpose, it is necessary that the state should transfer a part of its property and place it at the disposal of working people's teams, on a contract basis, applying leasing principles. Leasing ought to be regarded as an excellent way to implement principles of complete cost-accounting within the frame of state ownership.

A deeper stage of de-nationalization ought to be consistently implemented by transforming state-owned enterprises, those on lease being among the first on the priority list, into public share-holding associations and companies. The state's profits in this case will not only be taxes, but also dividends. A possibility exists for teams of the people working at these enterprises to buy them out, thus turning it from a leased unit into a collective property unit.

The above mentioned directions and ways of de-nationalization would first of all involve enterprises running at a loss. A task ought to be set to de-nationalize these enterprises by the beginning of 1991 in industry and in 1992 in agriculture.

State enterprises that will continue to function as such and thus be managed through administrative methods may include power stations, all large-scale enterprises of the national electricity grid, rail and road transport, aviation and merchant shipping, communication and defence-oriented enterprises and certain others to be listed in a special piece of USSR legislation.

Workers at such enterprises will be employed by the state and be guaranteed a sufficiently high level of payment for satisfactory work, along with lodging facilities, public health and other social services.

Of utmost importance is the establishment and guarantee of diversity of forms of property ownership in agriculture. Support of every kind must be given urgently to the development of family farms, along with small agricultural cooperatives. Land must be leased to citizens willing to take it, and collective and state farms running currently at heavy losses must be dissolved. Farmers can either stay in collective farms or leave them and get the plot of land due to them, set up a cooperative, a workers' association or homestead farm of their own. These provisions will foster equality of rights, under the philosophy of self supporting enterprises paying at the market rate for the use of land.

4.2. Restructuring of Financial-Crediting System, Financial Recovery and Money Supply Stabilization

In order to ensure the reform is truly radical, one of the top priority tasks is a recovery of finance and stabilization of money supply. With this in view a whole complex of steps and most urgent and extraordinary measures is absolutely necessary within the next two years, that would be able through the application of the reform principles to stop the situation getting worse. In this respect, along with dissolving and closing state-owned enterprises, it is expedient to consider questions like introducing parallel currency, a temporary taxation on production investments, mass-scale transformation of state enterprises into share-holding companies through buying shares of respective enterprises with the money accumulated and saved by the workers at those enterprises.

In order to decrease pressure on the money accumulated by the population, it is necessary widely to encourage savings. It would be advisable to introduce a higher interest rate on all kinds of "term-investments" in the USSR Savings Bank, to increase considerably the share of cooperative housing construction (covering up to 65–70 percent of all new apartments), to sell off some state-owned housing. Surplus cash could also be mopped up by sales of country houses with land, by raising special purpose loans from the population which would guarantee certain living facilities in return, and by encouraging purchases of durable consumer goods and equipment of all sorts.

Decisive for financial recovery is undoubtedly increased labor productivity on the basis of expanding and introducing on the widest possible scale, cost-accounting and economic self-sufficiency. Introducing a uniform principle of taxation and encouraging credits would be an extremely important step on this way. Performance standards will no longer be deduced for individual enterprises, but, on the contrary, all enterprises will have to try to reach a common standard of efficiency.

New legislation on the USSR State Bank and other banks will be the center of the credit reform. The present credit system mostly performs monitoring and supervisory functions, including distribution of the scarce resources. It ought to be replaced by a modern two-level banking system, the lower level being commercial banks, working on the principles of complete self-sufficiency, employing independently all the credit resources they can find. The upper level is manifested by the USSR State Bank performing the functions of a central bank, to monitor the functioning of the whole system of banks through setting interest rates, auditing the performance of commercial banks and of other banking operations. This will ensure a more reliable supervision of the money supply and its movement than at present.

It is necessary to increase the interest paid on deposit accounts thus preventing through creating an economic barrier an irrational approach to the use of economic resource. It is especially important keeping in mind the elimination of fees for production funds. Respectively, interest on loans ought to be increased to limit demand for credits.

4.3. Wage Reform, Labor Relations and Social Support of the Population

The existing situation in the society at present features a rapid polarization of incomes and emerging of contrasting differences resulting in an ever greater social-psychological tension. Taking into account the foreseen development of the market economy radical changes are urgent in the system of wages and in labor relations, as well as development and introduction of a special system of social security.

The major aim of wage reform is increased flexibility, and ability to adapt to changing conditions. All working people ought to have social guarantees of minimum wages, depending on their professional qualifications to a certain extent. The state system of tariff payment in wages will change radically. The new system ought to define standards of salaries and wages differentiating mostly on the basis of qualifications (skill and experience) with elimination of differential among trades. Conditions of labor, including adjustments to compensate for climatic, territorial, hazardous and other conditions ought to be standardized to a national system of bonus payments.

The total income of any individual worker, made up of his or her wage or salary, shared profit, investment interest, etc., can only be monitored through taxes and ought not to be limited in any way.

A comprehensive improvement of labor law is necessary, with adaptation to the conditions of the socialist labor market. Conditions of employment, dismissal, rights of payment, working hours, etc. are all involved.

A new mechanism of pension insurance ought to elaborated, too, which, based on the minimal living standard would enable all aged and disabled members of society to maintain their level of life with pensions and reliefs being indexed to the cost of living.

Development of market relations will result in worsening unemployment. The state will have to face a considerable expenditure on re-training, vocational training and reallocation of the work force. It will have to allot temporary unemployment relief payments. It would be sensible to use at least a part of these allocations for help in kind to the unemployed and their families. Creating such funds cannot be postponed any longer.

Social support for of the population also includes measures aimed at compensation for rising prices. No further steps can be taken in exposing pricing to market forces without first making provisions for the above. Such provisions could facilitate or hamper introduction of a market economy. They ought to include: determining the cost of living nationally, for individual regions and for areas, for various sectors of the population; determining the basic cost of living; elaboration of an assessment mechanism for the cost of living and introducing on this basis a system of population income indices.

4.4. Forming a Market, Restructuring of Planning and Price-Setting

The next five or six years should be the time when such closely coordinated and mutually dependent questions as creating a market, elaborating principles of planning and price-setting are solved. This will require introducing free prices for at least 5–10% of all goods produced over the pre-set state orders quota. In this way enterprises could use at least part of their accumulated means to start getting adjusted to the supply and demand market. In future the volume of produce distributed in this manner will certainly grow fast, up to 90–95% by 1995, as financial recovery proceeds and the trouble stabilizes.

After that the first priority questions will be those covering the efficiency of the market mechanism where principal efforts ought to be made to develop a competitive economy, with competition among manufacturers. Specialized antimonopoly programmes will have to be elaborated and implemented, involving antimonopoly legislation, developing a network of small-scale enterprises and of specialty production at existing enterprises, dissolving organizational monopolies of many types, including associations, research and development institutes, head departments in some industries that unite a major part of the national and/or republican markets. Stronger foreign economic ties will be essential, with a considerable opening-up of Soviet economy generally. For large enterprises embracing a major part of the home markets an obligatory exporting of at least a part of their product ought to be introduced, with failure to so attracting antimonopoly sanctions. Customs duties advantages, on the contrary, ought to encourage the import of goods and products into the home market which is otherwise monopolized.

Restructuring and planning must be directed to solve a dual task. New methodology in planning, with emphasis on the long term, but also programme and goal-oriented must be implemented which will proceed from the fact of the existing market. Based on market information, regulation of development in step with market developments will be the goal. The thirteenth five-year plan period offers an extremely favorable opportunity for this, which ought not to be wasted in any respect.

The plan for the thirteenth five year-plan period ought to be drawn as a plan of implementation of radical economic reform. The following must be ensured: distribution of state investments according to the requirements of society and the economy, stabilization of the rouble, development of production and market infrastructure. The plan also ought to determine a realistic relationship of prices, wages and salaries and other income of the population with respect to the years of the period coordinated with the state financial balance, it ought to foresee the necessary measures and reserves in order to ensure the regulation of these parameters.

Secondly, during the transition period, planning ought to complement and monitor the development of the market and through the procedure placing state orders. Mandatory state orders would be placed only at state enterprises while other enterprises may fulfil them on a contract basis. Economic methods stimulating the attractions to an enterprise of a state order may be elaborated, including flexible pricing practices.

Transformation of pricing policy is the key to reform in the very near future. Depending on how it will be effected, financial recovery, market formation and growth will reveal themselves, leading eventually to a balanced economy.

It, however, cannot be confined to a one-time routine revision of prices, since it will not then be the principal, vital factor ensuring the action of market mechanisms. Such a measure will fail to breathe life into prices to make them flexible, responsible. Present conditions seem more suitable for a reform of price-formation implemented as a stage-by-stage process of transition to the dynamic system of prices that would balance demand and supply.

The following order of implementation is possible: three types of prices are introduced, namely fixed prices determined by the USSR Council of Ministers, covering chiefly primary resources; regulated prices (limiting prices) determined by state bodies placing state orders including cases of contracts for state orders; and free prices (contract prices) depending on a contract between the parties involved.

Fixed wholesale prices in raw material industries will be increased in a closely coordinated manner over the period of four or five years starting from 1990 consistently directing them towards world prices for raw materials and energy.

The regulated wholesale prices for the products of manufacturing, process-ing and other industries, change depending on the requirements imposed by steady transition to open trading. Dual prices are possible at a certain stage, one price for a state order, the other price for the same product distributed freely through direct agreements with consumers or bodies and agencies of the Gossnab (this practice has been introduced already).

A strategy for price reform in agriculture is central fixing by local authorities of the system of rental payments eliminating a direct interference of the state in determining price levels of particular types of product. Creating the system of assessing natural resources and planning the scale of rental payments will be a vital and urgent problem for republican and local authorities.

Socially the most complicated problems will be connected with the reform of retail prices and the restructuring of the consumers' market. It should be admitted that under the prevailing conditions of a huge budget deficit, any reform of wholesale and purchasing prices, and formation of a market as such, make successful implementation of economic reform impossible if the existing retail prices and pricing policies are kept. Flexible prices responsive to the dynamics of supply and demand will promote a balanced consumers' market, and, consequently, will overcome deficiencies in the economy, black marketeering, profiteering and the black economy. Sooner or later we will have to face the necessity to introduce painful retail price increases.

A system of social support for the population will make increases acceptable. Retail prices can sensibly be restructured and revised in the following way:
– introduce free prices of supply and demand on luxury articles, on the major bulk of imported goods, on high-quality and delicatessen foods;
– differentiate fixed state prices on staple foods depending on their quality, preserving today's prices of low-grade foods;
– endow republican and local authorities with the right to regulate prices for goods requiring state subsidies, and with grants to cover those additional expenses. They will have the right to take account of public opinion to solve the question of a proper use of the money themselves. In order to achieve a balanced market within the frame of the available resources, local bodies should have the right to introduce ration cards, coupons and other ways and means including limited distribution of foods.

A rigid financial and credit policy, development of the market and competition will allow economic factors to stop the rise of prices, while the new incentives for increased production, increasing diversity, quantity and quality of goods and services created by the new mechanism of economic management will in the long run ensure increased welfare of all the working people. There is no, simply no other way of solving this problem, whoever claims to be able to do so, by whatever means.

The improved consumers' market ought to be used for a total restructuring of the system of wholesale and retail trade in consumer goods. Today a rigid system of distribution of state funds unchallengeably reigns here, and this is inevitable in conditions of high deficiencies and fixed retail prices. The very system is a source of unearned incomes, unjustified privileges, economic criminality. The system should be done away with as one of the most dangerous monopolies. It should be ousted by a normal standard system of wholesale and retail trade through a network of enterprises in various forms of ownership.

4.5. Organization Structures: Transformations

A transition from the administrative-authoritarian management system to the market economy demands a basic transformation of organizational structures and patterns both in production and management. From the hierarchy of administrative subordination penetrating the whole of the economy and society from the top to the very bottom we are to make a transfer to a system of organizational relations where horizontal ties dominate, reflecting partnerships of equal right.

Experience has shown that transformation of organizational structure ought to involve provision of more stable and reliable conditions of functioning with as few reorganizational steps as possible, that remark being also true with regard to the central economic bodies, ministries and departments. Organizational transformations ought to be implemented when they are a natural consequence of actual changes in the distribution of functions, when those latter are transferred from one unit to another. These changes will primarily occur in connection with enterprises leaving the umbrella and domination of their respective ministries and starting anew as a new form of property, be it public, cooperative or any other; when the market is established, the volume of state orders is limited and funds become even more limited, with further consolidation of Soviet power at all levels. In the long run the ministries will only have under their subordination and supervision state-owned enterprises.

An optimal combination of large-, medium- and small-size enterprises is to be established, dissolving the current organizational monopolies. At the same time, large administratively independent economic formations of a new type including corporations, companies and consortia will emerge. They ought to be initially structured so that they can never turn into monopoly manufacturers.

During the transition period the work of the state bodies of management at all levels ought to be reformed in order to overcome the practice which has existed for many years where functions of state management were all mixed up. The questions reserved by legislation to be solved at the enterprise level should be exempted from the jurisdiction of the ministries. All functions that can be rendered or performed by contract groups and organizations ought to be regarded as contract services and thus ought to be excluded from the duties of state bodies.

The application of the industrial or departmental principle of structuring an organization administratively, with a hierarchical structure of enterprises, will make it possible to implement the economic independence of Union republics, to develop normal relations with and between territorial bodies of management and enterprises, providing everybody's rights are legally observed and protected.

When solving the questions of regional economy, management options should be avoided where administrative duties are given to republican and local bodies of authority in order to manage the work of an enterprise in a way akin to the previously condemned practice of centrally managed enterprises. Relations between enterprises and territorial administration bodies ought, as a rule, to be based not on administrative subordination but on norms and standards set by the legislature, where the rights and duties of enterprises are determined with reference to the financial basis and social structure of the republic and the region and the use of natural and labor resources of the locality. In order to ensure preservation and development of the original culture of the peoples and nationalities inhabiting our country (in the republics, autonomous, and other territorial-national formations) maximal independence in solving economic and social tasks ought to be given to them.

4.6. Restructuring of Foreign Economic Activities

Economic and political considerations both necessitate more significant foreign economic ties in the development of the Soviet economy. These economic activities ought to turn from a means to patch up occasional deficiencies in material balance sheets into a powerful factor of production stimulation on the basis of consistent integration of the national economy into the world economy.

The principal means to perfect the management of foreign economic relations is a further expansion and enhancement of independence of enterprises ensuring free export of all products they manufacture above the state order quotas; the right to use currency profits (including the right to open accounts in foreign banks and employ other forms of foreign investment).

Restructuring of foreign economic activities as a part of the total scope of economic reform suggests eliminating the system of differentiated currency deductions and introducing a single taxation system of foreign currency incomes and profits, of businesses and of individual people.

As the mechanism of market regulation of the economy develops, the necessity of a convertible rouble will gradually emerge. This will require import liberalization through regularly held currency auctions. A currency stock exchange where foreign currency resources of enterprises and the ever-growing stock of currency centrally accumulated by the state could be traded, might even be set up.

An active search for new sources of material and financial resources will be necessary to ensure the steps enumerated above, one of the top priority measures being a special programme of attracting foreign investment to the USSR.

5. Stages of the Reform

Carrying out radical economic reform makes big demands on management organization of the transformation processes. We ought to keep in mind all possible consequences and the duration of the processes and prepare steps well in advance in case various situations may arise, to assure a consequential and consistent progress along the way outlined.

The logical succession of the reform measures and steps arises from the necessity to achieve a real surge in the economy which can only be guaranteed through a substantial intensification of incentives in labor and economic activity, as well as through an improved coordination of efforts of all participants in the social production process.

The *first* stage is a preparatory set of measures effected on a one-off basis to create the economic mechanism of the transition period. It is to be effected in 1990.

At this stage legislation is worked out and laws adopted to give an impetus to the transformation of property and ownership relations. These include a Law on a united taxation system, and a Law of the USSR State Bank, whose primary function is to promote cost-accounting and make a step towards financial recovery.

During this period all processes ought to be activated to accelerate and encourage transition of enterprises to leasing, their transformation into cooperatives, public share-holding associations and other economic formations.

Reforms of pricing, wages and salaries, and social insurance are being elaborated. The system of indexing profits and incomes is being introduced, and corresponding pieces of legislation are being adopted. Regulated prices in wholesale trade can be increased at this stage, and commodity auctions can be introduced and encouraged. Currency auctions are set up. Questions connected with reforms in planning are being closely considered. Personnel retraining is extensively carried out.

Extraordinary measures aimed at financial recovery are implemented, to reduce the budget deficit, to curb growth of incomes, and to limit credit. By the end of 1990 all enterprises running at a loss are to be closed or done away with in some other way. In industry they will be transformed into leasing enterprises, cooperatives, share-holding associations, etc. Facilities are developed to increase the production of consumer goods. The structure of imports is changed. Measures are taken to minimize losses and waste in agriculture. Legislation is being elaborated and adopted concerning economic independence of the Union republics, regional cost-accounting and self-sufficiency and on local self-government.

At this period a possibility of a slowdown in production and an increase of deficit is to be anticipated.

The *second* stage involves a complex of measures – mostly of a one-off nature resulting in launching of the economic mechanism of the transition period. The stage embraces the period 1991–1992.

All the above-mentioned legislation is brought to life, steps are taken to ensure their enforcement. The indexing system allows the start of pricing, wage and salary reforms. The new system of planning is adjusted, along with the system of placing state orders and application of regulated prices.

This stage should feature an accelerated tempo of de-nationalization in all spheres and branches of national economy. By the end of 1991 all enterprises running their business at a loss ought to be closed down in agriculture, and homestead farms ought to be encouraged on their sites, and cooperatives ought to be encouraged to open as well as other forms of business. Highly complicated processes of getting adjusted to the new taxation system will occur, and the new, more rigid policy would cause some complications, too. The USSR State Bank restructuring being completed, the credit reform will set in.

All this will entail a rapid development of the market in connection with permission to distribute the products of businesses in the market if they exceed the quota set by state order and if the latter has been met. Commodity auctions and regularly held trade fairs are transformed at this stage into stock and goods exchanges and commercial centers. The Stock Exchange is set up.

The second stage is a decisive period in launching the mechanism of a transition period. A break-through atmosphere can and must be created, manifested in the termination of the slowdown in production and the beginning of its recovery with consumer-oriented products in the market. First tangible results are to be clearly felt by this time of the attempt to increase the manufacturing of consumer goods. However, due to the incomplete financial recovery, with disproportions in production not yet overcome, a result of the monopoly character of the economy still to be replaced, the deficit is highly unlikely to be done away with by that time.

The *third* stage is the adjustment of the economic mechanism of the transition period, with implementation of the programme of the reform development. It is to embrace the period 1993–1995.

This stage of the reform will feature a complete financial recovery, with implementation of the antimonopoly programme, including elaboration of relevant legislation and the necessary restructuring of organizational patterns. Without the financial recovery in conditions of deficiency the antimonopoly struggle will be very ineffective.

Due to the expanded market sector and the policy of monitored inflation both prices and incomes will grow. Setting up a two-level banking system will exert pressure on these processes through credit limits, through interest-rate policy and other measures and steps. The state will acquire a powerful tool of curbing inflation.

By the end of this stage a balanced consumers' market will be achieved, a marked development of foreign economic relations will be reached, the climate for foreign investments will improve considerably and they will start paying off, thus contributing to the satisfaction of the market. Conditions shall be provided for a partially convertible rouble, probably through an application of the practice of parallel currencies. Competition will start playing its part.

A realistic assessment of the profundity of the crisis and of the possible and optimal rates of overcoming it makes one suggest that the third stage will most probably be a period of stabilizing of economy and ought not involve a high growth-rate. Investment programmes shall not be overstretched.

The *fourth* stage is a period of taking a firm foothold for the new economic system. It shall bring the final stage of formation of all related structures and patterns in the production sphere and in the sphere of socio-economic relations. It will cover the period of 1996–2000 and further on.

This stage will allow creation of strong and effective incentives of economic revival and growth and effective improvement of the people's well-being. If we manage to solve the tasks outlined above, the reform will bring its fruit in full measure, the fruit so much expected by all Soviet people.

Chapter II
Introducing a market economy in the USSR: pressing problems

Nikolai Y. Petrakov

Professor and Corresponding Member of the USSR Academy of Sciences, Assistant to the General Secretary of the CPSU

The overriding objective of the economic reforms underway in the USSR today is enhancing the efficiency of national economic management. To attain this objective we must rely most heavily on market methods in regulating and managing the manufacture and distribution of products. At the present time, efforts are being made to remove ideological dogmas that have barred rational economic thinking from shaping our economic practices. It is common knowledge that over the last seventy years the socialist model has undergone the most drastic modifications as regards the marriage of market forces and central planning.

Originally it was axiomatic for us to maintain that planning was the antithesis of market. It followed that market was to be destroyed with all of its functions taken over and performed successfully by fiat of centralized management. This speculative concept had been prevalent up until 1917 and could be dated back to the preMarxian period when utopian socialists entertained such views. Vladimir I. Lenin dealt those views a crushing blow. He examined the realities that obtained in Russia after the October 1917 revolution and the civil war and argued that market methods were what the Soviet economy needed badly throughout the period of transition, that is, up until socialism's definitive victory in the USSR. Vladimir I. Lenin's arguments were so sound and his scholarly and political authority was so overwhelming that Stalin could not dare to reject his concept. While placing harsh limitations on the scope of market relations, Stalin recognized a likely "marriage" of socialism and market in his theoretical studies but believed that the government should keep an exceedingly tight rein on such relations. Though having undergone a variety of modifications, this viewpoint dominated Soviet economic thinking until very recently. Now the situation has made an about-face. We now admit publicly that market-based management mechanisms are inherent in the socialist type of economy. More importantly, numerous socialist criteria, such as distribution according to one's labor, self-government, the freedom to choose what one likes to do, social guarantees propped up by decent consumption levels for man, are impossible to implement without massive reliance upon market regulators. Our awareness of this relationship has compelled us to study problems of how to put the Soviet economy on the market footing in earnest. There are a lot of hurdles lying ahead for us to clear by means of political will and a well-thought-out economic policy.

At this juncture, it is hard for us to introduce market relations and managerial practices based on economic incentives because our monetary system has, in fact, been eroded in our economy. Money, neglected as a relic of the prerevolutionary system, to be removed as soon as possible, is there to revenge itself. For many a year our managers have been led to believe that while Western entrepreneurs are busy "making money", we are expected to make machine tools, bricks and tubes under socialism (as if reaping profits does not call for managerial practices devised to meet a society's and an individual's requirements). As a result, money is the only "merchandise" that is not in short supply. And we are hard-pressed to part with our money in exchange for goods and services. Money keeps on piling up on factories' and individuals' bank accounts, money's turnover slows down, and people's purchasing power does fall off. A massive flight from money occurs and demand for non-perishable commodities rises by leaps and bounds.

Such an environment makes it impossible to take advantage of economic independence, to unfetter initiative and to provide incentives to boost business activity. The country's monetary system should be rendered healthy. The prescription is simple: surplus paper money should be taken out of circulation. So far we have taken no practical measures in this field. Instead, products and resources are rationed and allocated in quotas everywhere and capital investment is earmarked in prescribed amounts. As a result, we continue to veer away from a market economy governed by its own laws of distribution and to move back to the laws of our administrative system again. Our words are, again, not matched by our deeds. This practice results from our primitive approach to the economic accounting principles: you get your pay so long as you continue to work like you did before, that is, you do not work for a specific consumer of your product but rather for "your own" ministry and for the government order put together and placed by "upper echelons". The government does the planning, the manufacturing and most of the buying of what it manufactured. The end user, or, in other words, the population, collects its full share of pay and bonus money from the government but not the products and services the end user needs. As a result, workers' savings continue to pile up to the tune of almost 100 million roubles a day in the Savings Bank (they shot up by 17.7 billion roubles over the first six months of 1989). 8.9 billion roubles in paper money was printed over the first six months of 1989.[1] The more money we have in circulation, the farther we move away from a market economy.

Now it is crystal clear what measures we should take to render our national economy healthy in monetary and financial terms. Such measures ought to include: 1) drastic cuts in inefficient centralized government investment in the sphere of production. The scale of capital construction projects should be brought in line with the actual capacities of our construction industry with its structure adjusted to meet the requirements of efficient structural reforms in the national economy; 2) drastic increases (by 10 to 15 percentage points) in the proportion of consumer goods and services produced within the gross national product; 3) speeding up the rate of conversion of defence industries to peaceful occupations and continued reductions in army troops; 4) strong measures designed to divert the population's monetary savings from meeting current demand (into, for instance sales of apartments and summer homes, special-purpose loans to buy cars, etc., sales of stocks and bonds, higher interest rates for fixed-term deposits, etc.); 5) curtailing the almost automatic granting of credits for enterprises and other government practices to keep afloat enterprises with poor performance records.

Some initial steps have already been made in each area indicated above. One can argue whether or not the action taken was vigorous enough, or whether or not the steps made were consistent enough. But what is obvious is that time is needed before tangible results will ensue. And a lot of time is needed. For example, even if we, all of us, throw our weight behind the effort, our adjusted structural policy will bear fruit only in about six to eight years from now (if one is an optimist). Other measures will yield palpable results (that is, results that will affect the entire national economy) in three to five years from now. Is that not too long a time? Do we have so much time to spare in the light of heightened social tensions, people's shortness of temper fuelled by shortages of goods and by the introduction of coupons to meet our basic and not so basic necessities as well as the flourishing of a black economy in the country?

I do not think so. There are at least two problems that should be handled without delay, for if we start these engines, we will be able to steer the country through the quagmire of economic chaos and put it on the firm ground of healthy economic development. The first problem is related to the need for an environment that will be conducive to invigorating the business activity of economic units and all types of enterprises. The second problem involves a change-over to the open economy concept, that is, the Soviet Union's integration into the international division of labor system on a modern footing (or on a civilized footing, civilized being an in word now). If we fail to live up to a host of economic and technological requirements presented by the world market,

---

[1] Izvestia national daily. September 9, 1989, p.1.

we will be unable to introduce the brand new technologies we need for the production of goods necessary to gratify people's needs. Accepting the "ground rules" of the world monetary and financial system is not surrendering to the mercies of capitalist market. On the contrary, as long as we remain on the sidelines of world economic ties, we suffer great economic and political losses.

A solution to the first problem calls for a monetary reform, for in the absence of a steady monetary unit we will be hard pressed for business activity, initiative and efficient economic incentives. A solution to the second problem hinges on how to make the rouble convertible. Now everybody recognizes the need to render the Soviet money convertible. The press never tires of bandying the topic, setting one's teeth on edge. But we have been marking time recently. Some ritual dance has been performed, with the implication that it is nice to have one's own hard currency but we can have one only when we have reached the end of the tunnel, that is, after we will have completed a structural revamping of our national economy, putting it on a modern technological basis and building a highly competitive export sector. But are we to attain this objective by means of a forced changeover to total rationing, a system that will act as a running knot strangling the economic inducements we seek to employ in our managerial practices? How can we attain this objective in the absence of far-ranging scientific, technological and economic contacts with the world economy, for our evolving contacts are ensnared by the caveman's barter deals (oil is exchanged for machine tools, timber for stockings, etc.)?

One forms the impression of being caught in a vicious circle. The optical effect of impasse is produced for the simple reason that originally a seemingly perfect but organically defective scheme of logic was adopted: "first we do this and that, next we do this and that". Problems "queue up", as it were. The way out is to solve all the problems concurrently. Why should a solution to the problem of building export industries in the national economy be placed out of touch with the efforts at establishing a realistic exchange rate for the Soviet rouble? Why should structural economic reforms not rest on a "hard 10-rouble banknote" payable only for highly competitive products in high demand on domestic and international markets? We have many a reason to start introducing a money reform now without further delay and it should be coupled with efforts at rendering the rouble convertible.

Numerous developments that do occur in our national economy provide graphic evidence that hard and credible money is urgently needed. Some regions have recently been discussing the highly popular concept of introducing a viable republican currency. If one sets aside the political dimension of the issue, one will see for himself that such concepts stem from the lack of confidence in today's rouble, for too many roubles chase too few goods and the rouble's purchasing power varies from region to region, resulting in people roaming about the country looking for goods to buy.

Hard currency has been used as an incentive for our export producers for a fairly long time now. They may spend a part of their foreign exchange earnings as they see fit, buying raw materials, equipment and consumer goods. Dollar shots in the arm are the most graphic evidence we have, illustrating how inefficient it is to use "soft" roubles as an economic inducement.

That the conclusion is right has been corroborated by our recent decision that Soviet suppliers of high-grade wheat and some other crops should be paid in hard currency funds. This decision is as remarkable as it is naive in economic terms. It stems from what we call an import replacement concept: a domestic wheat grower is entitled to collect a part of the foreign currency savings made on imported wheat, his part being calculated as proportionate to his domestic contribution to imported wheat replacement. But it is the imported wheat replacement that may fail to materialize even if some collective and government farms manage to double, or treble their output. The point is that the country's grain output is made contingent largely on the average yield nationwide rather than on the performance of some individual "pace-setting" farms or even regions. Bumper crops in some farmland areas may be reduced to naught by droughts, or rains elsewhere in the country.

But this is not what really matters. This grain payment experiment is highly unfair to a great number of the remaining farmers and industrial manufacturers. Why is it that grain producers have gained such advantages? Have they produced enough food for the country? Or are we less eager to reduce the volume of imported beef, butter, light industry products, rolled steel, and equipment? And what is the financial status of those who work to pay for our imports today: workers employed in the oil-, gas- and timber-producing industries? Their lifestyles are a far cry from those enjoyed by Arab Emirate sheiks. Lastly, this experiment has provided yet one more proof that the avowed equality of various forms of ownership and employment has been given a cold shoulder. A farmer making his own lease-holding arrangements, or running his own farmstead finds himself "short-changed", for no matter how hard he tries, dollars will be collected only by collective and government farmers.

This is the host of problems and puzzling questions that the grain-for-foreign-money experiment raises. And there is but one way out of this poignant situation: everybody should be treated as equal and everybody should be entitled to collect pay in hard currency if he or she performs well. But it is the marketplace that should certify the quality of the work done, not the quality control inspector, or the government quality assurance inspection commission.

I believe that hard money is what a nascent market needs in this country. If market relations evolve on the basis of today's inflation-ridden rouble, they will result in sky-rocketing prices, in a market dominated by government-owned manufacturing monopolists and by individuals and groups of people possessing immense financial and property assets that are, as a rule, not legitimate earnings as well as in a massive demand that wages and salaries should be raised to make up for soaring prices, not for better work. There is no future in such a market. It will fail to encourage the drive for a higher quality and competitiveness because right from the outset it will collapse under the burden of problems and imbalances held over from the times of management by fiat and domination of arbitrary decisions and ambitions projects devised to make headlines.

Well, a monetary reform is what we need as badly as the air we breathe. But how is it to be effected? Overnight, within a rigidly fixed time-frame, or bit by bit? History supplies a variety of examples. For instance, West Germany effected a "blitz" reform of the German mark in 1948. A few months prior to that reform the USSR took an action that was similar in execution but somewhat different in form. But the USSR is a country where another and more brilliant monetary reform took place: a gold 10-rouble coin was minted and circulated in 1922–1924. This version of a "soft monetary reform" is better suited to today's economic situation in the country. I would like to attempt to adduce arguments to prove my point.

First of all, you will recall that we need a credible monetary system as a means for economic recovery rather than as an end in itself. Credible money should be made to cater to real market relations. Realistically speaking, market relations are unlikely to dominate our entire national economy overnight. A period of transition will certainly be needed. If "new" money is introduced everywhere at the same time, the unwieldy State sector with its inertia-driven system of management by fiat will immediately generate the supply of credit money and the profits accruing from our monetary reform will prove short-lived and vanish into thin air in no time. And the profits accruing from a monetary reform in the longer term are not an overnight success in eliminating surplus money (such surplus paper money will start cropping up as long as our economic system's structural and managerial "headaches" remain). Monetary reform will accomplish its strategic mission only when our domestic money establishes durable links with the world's monetary system. And we are unlikely to do that throughout the "space of this country's economic life". But we will be ill-advised to allow any delays.

It follows that a convertible rouble, or a hard rouble, should at first be made to cater only to the operations of joint ventures, open-sector enterprises (that is, a sector that will operate in a world market mode), and enterprises that turn out export-oriented, or prospectively competitive products (domestic products designed to replace imported ones). Such a rouble should have a realistic exchange rate as regards ECU and major world currencies. To demonstrate that its exchange rate is realistic the hard rouble should be exchanged freely for other hard currencies. Fundamentally, some other country's currency allowed to circulate freely in Soviet territory may play the role of such a hard rouble. The term "free circulation" is used in a broad sense in this context, even though it may, as a matter of fact, be applicable only to dealings among enterprises, or between enterprises and the government, that is, the hard rouble may be barred from dealings among private individuals. Properly speaking, the grain experiment lays a strong groundwork for future events to take such a course. It appears to me that such a course is less preferable than one devised to render the rouble healthy through its organic integration into the world monetary system.

It stands to reason that the rouble may only acquire a realistic exchange rate if a domestic hard money market is established. To this end fundamental changes should be introduced in our system of distributing a relatively small amount of the petrodollars this country earns. Instead of splitting this foreign money "jackpot" on the "roll-call" basis, arrangements should be made to start selling convertible funds to Soviet enterprises and organizations at a free market exchange rate. The foreign money auctions everybody hears so much about may be viewed only as a warming-up before the go signal is given. The auctions serves only as a training aid for the bidders and viewers. But a real exchange rate for a currency is shaped by the play of market forces that are not limited to the 50 to 60 million hard roubles' worth of funds offered for auction by managers and executives indulging in their whims.

A "soft" monetary reform is attractive in that unlike numerous roubles of Tambov, Khabarovsk, Moscow mintage, etc., a convertible rouble will not only have a valuable appearance but, most of all, will be freely exchangeable for Western currencies at a standing exchange rate and will not be planted in our economy by fiat but will force its way, as it were, into the sectors where only convertible funds are of use. For instance, a part of the profit reaped by the Western partner in a joint venture on the domestic Soviet market may be repatriated in hard money, or additional investment to promote expansion in joint ventures may be financed by convertible funds. Lastly, hard money will be needed to finance transactions between the export sector and the rest of the national economy.

If our national economy is put on the track of market relations, the exchange rate of our hard rouble will keep steady "on the right" and "on the left", that is, in relation to Western currencies and in relation to our "podgy" rouble now. This will mean full recovery for the latter rouble and a completed monetary reform for the national economy. If the exchange rate of our "podgy" rouble continues to decline in relation to our hard rouble, two alternatives will enter the picture: if the hard rouble's exchange rates keeps steady "on the right" (that is, in relation to Western currencies), it will emerge dominating the domestic market. If, God forbid, things go wrong and our hard rouble begins to "float" vis-a-vis Western currencies, our economic reform will go down the drain and we will see our old centralized supply system alive and kicking again and welcome a new rationing policy.

It follows from what we discussed above that, when introduced into circulation, our hard rouble will not relieve our government departments of the need for a strong policy to render our national economy healthy in monetary and financial terms. But to do that the domestic exchange rate of our "soft" rouble should be stabilized as a matter of priority.

The grave deficit that erodes our budget stands in the way of efforts at stabilizing our domestic monetary unit.

Curtailing the government's inefficient spending is a major weapon for obviating the national budget deficit. Of course, efforts to reduce our defence spending and to convert our defence industries to civilian output are a crucial guideline for us to follow in this area. But we also need to overhaul our investment policy countrywide. Uncompleted construction projects went up in cost by over 30 billion roubles from 1985 to 1988. This signals an increase in financial investment in construction projects that feature protracted construction and cost recovery schedules, scattering expenditures to finance too many construction projects, and longer construction schedules for each project. About 7 billion roubles in cash overhand was paid to construction workers as wages for uncompleted construction projects over the last three years. We contend that the amount of government investment should be reduced drastically. The volume of construction financing must be brought in conformity with the construction industry's actual production capacities. Moth-balling construction sites ought to be viewed as a major government programme implemented under the auspices of the USSR Council of Ministers and central planing bodies because moth-balling calls for a well-coordinated policy to be conducted at every technological stage, ranging from raw materials to finished products. It is beyond individual sectoral ministries' power to do the job.

Our investment policy needs decentralizing and ought to be guided by economic efficiency beacons rather than by the administrative ambitions of governmental departments and planning bodies. The Soviet economy is confronted with the need to introduce a capital market. A reformed banking system we have now begun to develop is likely to introduce competition in the field of capital loans. A network of commercial banks, including cooperative banks, will encourage business initiative, reasonable risk-taking and business rivalry.

Plural investment activities will also be promoted by the massive spread of joint-stock systems of concentration and redistribution of financial resources. Now relevant legislative acts are being finalized to float Soviet joint-stock companies. Joint-stock systems have a number of advantages in that the national economic management system may be rendered more mobile. Firstly, joint-stock companies and societies help remove the bad practice of solving economic problems at government expense. Secondly, financial resources allocated for investment gain in mobility. And thirdly, real democratic criteria begin to shape

investment policy. Joint-stock companies invest their money in capital construction projects by means of voluntary commercial contributions to establishing
and shaping economic ties and contacts. The acquisition by a government-run
enterprise of another enterprise's stock (and paying for the acquired stock with
its own growth fund money) is a unique form of voting, where a team of workers
passes a vote of confidence in specific economic undertakings and pioneering
projects. While purchasing, or refusing to purchase other socialist enterprises'
stock, companies and associations make their de facto identification of economic
priorities in their society.

It follows that the launching and expansion of socialist enterprises by way
of joint-stock ownership offer a means to obviate the monopolistic investment
practices of the central management body that played a substantial role in
promoting major structural shifts but slowed down the country's economic
development in recent years. The overcentralized capital investment that goes
beyond the objective needs of economic development paves the way for non-
economic considerations to dominate capital investment policies and amounts.

Our new approach to the entire price-setting system is crucial to the efforts
at introducing a socialist marketplace. In this domain we will have to jettison the
stereotypes of our traditional economic thinking that have been fostered by the
ideologists advocating rigidly centralized economic management practices.

In order to set up a socialist marketplace, we need to make consistent efforts
at renouncing manufacturers' monopoly, discarding the "coupon" system for
distributing resources, reinstating the rouble as a universal equivalent, vesting
enterprises with real independence in choosing their own economic partners
and with full responsibility for their performance vis-a-vis the consumer and
the government. This is how we view the perspective and meaning of our price
reform that ought to bring about a flexible price-setting machinery that will
respond without delay to the outcome of interaction between demand and
supply.

Our prices and price-setting machinery may be reformed through two
alternatives: the first scenario provides for a concurrent and overall revision
of wholesale, purchasing and retail prices and the second scenario envisages a
stage-by-stage across-the-board revision of all types of prices accompanied by
appropriate measures taken at every stage to place enterprises' and people's
growing income under strict control, to dismantle the current system for the
quota distribution of physical resources, to reduce the government's inefficient
expenditure, to improve the country's investment policy, to overhaul the taxation
system, and to enhance the banks' authority in bringing money circulation back
to normal.

We will examine in brief the advantages and disadvantages of each scenario.
The advantage (perhaps, the only one) of such a one-time global revision of all
prices lies in the hypothetical likelihood that we will be able to do away with all

the bad vestiges of the past at one swoop, to make the prices mirror the real state of affairs that obtains in the national economy as a result of our structural policy in recent decades, and to set a formal stage for sectors to proceed to self-financing, that is, to clear perestroika's construction site of outdated prices.

But we maintain that it is not feasible to take advantage of this likelihood for the following reasons:

– a total revision of prices for millions of articles with due regard for the real production costs and efficient use of each article is impossible. A correct price reform concept developed by the upper echelons is inevitably transformed into routine errors that are made when a tremendous number of specific prices for specific articles are calculated;

– a one-time revision of prices is always worked out on the basis of the tidal wave concept, a wave that rolls from commodity-producing sectors to consumer goods and services. A centralized price reform is essentially a series of overt or covert compensatory payments to the consumer for the rising prices of suppliers' products. But such compensatory payments do, in fact, rule out the use of prices as a means of encouraging the efforts to save resources, to apply more exacting procedures for selecting capital investment options, and to balance demand and supply in wholesale trade and on consumer goods markets;

– a price reform undertaken as a one-time governmental action against the backdrop of today's economic realities will be tantamount to "jumping the gun", for the general environment for a nascent socialist market mentioned earlier is very slow in coming along. A one-time price reform will slow down rather than speed up the advent of such a market. The idea behind setting market forces in motion is above all "invigorating" prices that should be rendered flexible and dynamic. A one-time price reform is actually a change-over from one set of frozen prices to another;

– and, lastly, a one-time retail reform, if reduced to massive and substantial price hikes, appears unacceptable for social reasons and will not yield any economic profits in terms of a healthier national economy.

In this connection the second price reform scenario devised as a stage-by-stage change-over to flexible and dynamic price-setting procedures appears more attractive.

According to this scenario:

– the wholesale prices charged by commodity-producing industries will rise through coordinated action within 4 to 5 years, starting from 1990, so that they could reach gradually the level and correlation of world prices for commodities and energy resources. The bulk of such price hikes should be completed late in 1995;

– wholesale prices for the products turned out by machine-building and other manufacturing industries will be regulated by the needs of a consistent change-over to wholesale trading. Dual prices are likely to be charged for the same

products at some point in time: one price for the products sold freely by agreement with relevant consumers, or State Procurement Committee bodies (the latter type of sales have already begun). Free prices should be charged liberally for sales of scientific and technological innovations and for merchandise sold to meet enterprises' demand paid for out of a given enterprise's growth fund money. The USSR State Procurement Committee ought to progress from arranging wholesale fairs to commodity exchanges and auctions;
– purchasing prices should be set only for the basic products of farming and animal husbandry in the smallest possible number of specialized zones. The strategic thrust of a reformed price-setting machinery in agriculture is a planned centralized system of rent payments without the government's immediate interference in setting price brackets for specific products. Given the population's surplus money a change-over to flexible and dynamic retail prices will add legitimacy to the inflationary processes underway in the country. This is to say that the government ought to abandon individual practical anti-inflationary measures and proceed to a policy of moderate and manageable inflation. Such a change-over entails, in particular, some substantial modifications to be introduced in regulating the population's wages, salaries and income. The government ought to regulate only minimal wage and salary brackets. Such minimal wages and salaries should be adjusted every year according to the country's retail price index. A similar procedure should be applied for retirement pensions, scholarships, and various social allowances and benefits.

The system of wholesale and retail trade in consumer goods stands in need of drastic reform. The system is one of rigidly centralized distribution of government-set quotas rather than one of trading. Stores and public catering facilities are barred from showing their own initiative, for they are hostage to municipal, regional and republic trading organizations that possess their own wholesale facilities. This system ought to be dismantled as one of the most dangerous monopolies. Selling on commission should be a criterion to be followed by our reformed retail stores. Trading firms, associations and stores will be licensed by the government to purchase specified products right from manufacturers, regardless of who owns them and who they report to. Such licenses will specify a minimal line of products to be sold by a trading facility. Its sales personnel will be paid out of the money deducted from the facility's actual volume of sales. The permissible levels of mark-ups on government-set wholesale prices (say, 5, 20 or 30 percent, or no ceiling at all) should be set for various groups of merchandise. Regular and small-time wholesalers may operate only on a profit and loss basis, that is, they will make a living by rendering voluntary services to the retailers.

Chapter III
The economic system of the USSR

Stanislav S. Shatalin

Member of the USSR Academy of Sciences

1. Laws Governing the Economy of the USSR

For a long time society has been organized according to the precepts of socialism, which have been viewed as invariable and presented as dogmas leaving no room for their scientific analysis. The critical study of real economic laws has been replaced by laudatory statements about the advantage of the existing system.

That system was established in this country in the early 1930s. The primary task then was forced industrialization. The creation of modern industry on the basis of an unskilled and poorly mechanized labor force ensured a leap in the volume of industrial production. The basic features of the management system under those conditions were:

– an hierarchical structure of the organs managing the economy, with complete administrative subordination of primary industrial units to the national economic level;

– the practice of empowering higher management echelons to define the objectives of development, the ways of achieving them, and the day to day coordination of activities of the subordinated economic units. This was done by working out natural balances of supply and demand, devising distribution plans, passing these decisions through the hierarchy to the notice of enterprises in the form of orders to perform tasks, to make production deliveries and to fulfil quotas.

Those characteristic features determined the whole structure of the economic mechanism. In such a situation economic stimulation for the fulfilment of the plans is quite natural. Directive assignments immediately couple the requirements of society to the economic system and a failure of a link in that system to fulfil them may bring about a chain of disruptions in the national economy. The dominating role of these hierarchical links, of directives coming down from the upper echelons and reports on their fulfilment going up create an illusion of rigid regulation and effective management of economic life.

The activities of enterprises are regulated by dozens of directive indices and hundreds of thousands of normative acts. Employees' wages are regulated by a system of tariff rates and salary rates, by regulations on bonuses. The establishment of prices for the main types of produce is a prerogative of state organs of pricing, while the rights of enterprises in this respect are extremely limited.

One of the arguments most widely used by opponents of serious measures aimed at expanding the economic independence of enterprises is that manage-ability of the economy might be threatened.

But the effectiveness of centralized management is to be assessed not by the number of indices and quotas sent down but by its ability to formulate a substantiated strategy of socio-economic development and unless this strategy is fulfilled, the notion of our economy's manageability has to be radically amended.

For instance, comparison of the targets of the approved five-year and yearly plans with the results reported reveals sharp deviation of the real dynamics from the planned targets not only in the production of some types of goods, but also when macroeconomic parameters are concerned. It is noteworthy that in many cases the real state of affairs directly contradicts the planned targets.

The process of accelerated growth of circulating assets in commodity stocks – material values which directly contradicted the targets of five-year and yearly plans – influenced greatly the economic development in the 1970s and early 1980s. More than once it has been decided to limit the volume of unfinished construction work, to correlate them with the volume of capital investments, and to concentrate resources on specific construction projects. Nevertheless, further expansion of the construction sector has taken place, with resources being scattered among many projects.

There is widespread conviction that poor management of economic pro-cesses has manifested itself only in recent years and that it demonstrates the imperfection of economic conditions, while in the years of formation of the national economy under conditions of extensive development high manageability was ensured. In our opinion, it is not exactly so. Of course, when the economy was not so complicated, the degree of real control on the part of the center was greater. But in that period as well the dynamics of the national economy used to deviate considerably from the plan, while those deviations in many cases led to rather painful consequences.

For instance, the first five-year plan envisaged that a sharp increase in the volume of production would be accompanied with an increase in real wages. But the forced growth of the volume of production demanded substantial redistribution of resources in favor of the accumulation fund.

The problem of supplying the population with vital consumer goods worsened considerably. The forced consequence of that was the development of a many-channeled, regulated retail trade with a considerable difference in prices depending on the channel and a considerable rift between market and state prices. As a result, state retail prices had to rise steeply.

The problems connected with the scattering of capital investment, and the growing scale of incomplete construction were already fully manifested at the beginning of the 1930s. At that time decisions were taken on concentrating resources, and on accelerating the pace of construction, which exerted only limited and brief influence on the real situation.

As early as the beginning of the 1920s Soviet economic literature became aware of the fact that centralized management and the regulation of economic activities were not the same thing, and that in fact the latter may conceal spontaneous uncontrolled processes. For instance, when analyzing the methods of economic management formed under the conditions of War Communism even such a proponent of it as L. Kritzman noted: "This system is organized only formally ..., in fact it is anarchic and its anarchy is an anarchy of supply ... when the goods received partially (and to quite a great extent) cannot be used and remain non-used". [1] But even today one can come across relapses into considering socialist economy as a relatively simple, static and determined system in which the center collects the information on resources and demands, works out a plan envisaging the optimal distribution of the resources available and ensures its fulfilment by issuing to the enterprises assignments of work taking into account all the reserves available.

The real economy is different from that ideal. It is a dynamic non-determined system. Consumer demand is constantly changing and it is beyond any exact forecasts and rigid regulation. Weather conditions greatly influence the results of agricultural production, the work of the industrial branches connected with it and the system of financial balances. Foreign trade conditions change dramatically and often unpredictably, the same is true of prices for major imported and exported goods. It is impossible strictly to regulate the results of scientific and technological progress; plans and forecasts here are inevitably of a probabilistic nature. The information on the real state of affairs in economics, on demands and resources is limited. The major factor causing uncertainty is the lack of precise correspondence of personal and collective interests with social interests, the impossibility of a strict control of the operation of the sum total of all economic subjects. Indeterminate does not mean impossible to plan. The fact of the matter is that in order to have more or less realistic plans one must work them out using methods which allow one to take into account the stochastic nature of economic development.

Whether or not management is efficient depends on the degree of power motivation, the desire to retain the post held and for a successful career. The personal interest in retaining a post greatly influences the degree of loyalty of the subordinate employees to those above them. It would be wrong to limit the

---

[1] L. Kritzman. The New Economic Policy and Planned Distribution. Moscow: Gosizdat, 1922, p. 24 (in Russian).

incentives for retaining one's post simply to a higher income or to fringe benefits and perquisites. Social prestige, the creative nature of labor, etc., are of great importance, too.

The effectiveness of power motivation is not an arbitrarily varied parameter, it is limited by objective factors. Such a mechanism can only work effectively if the rate of turnover of managers is within permissible limits. When the rotation rate slows down, the directives of the higher organs of management inevitably become less effective. The other sanctions available within the mechanism of power motivation are only effective if there is a real threat of dismissal.

That mechanism must effect the directives coming from higher authority. But there is no reason to believe that meeting all demands is compatible with an acceptable regime of rotation of managers. For instance, suppose a failure to fulfil the planned introduction of new technology, or to complete a construction project incurred strict sanctions against executives – beginning with ministers and ending with heads of industrial enterprises: we would have to replace most executives every year which would not, obviously, accelerate scientific and technological progress or promote success in construction. The demands made of managers of different ranks are brought into proper correspondence with the options made available by the mechanism of power motivation in practice. A set of major parameters is inevitably formed, with a failure to fulfil them incurring the strictest sanctions, with a set of other assignments which are optional to a considerable degree.

Which of the most important parameters that are really controlled by higher authorities depends on the sector of the economy being considered, on the economic situation, and on the priorities of the economic center, but it is not arbitrary. For instance, exaggeration of recent achievements is natural in the command and administrative system of management. The most important factor here is that recent results are definite and can be clearly connected to particular executives. There may be failures in the long term, such as lagging behind the best world standards in technology and production quality, exhausting soil fertility, damaging the environment, etc., but these are revealed only after the lapse of time, so it is difficult to connect them with particular persons who should be penalized. That is why the system of impermanent economic executives is gaining in strength.

Similar factors have led to a definition of production growth rates as being more important than the reduction of costs. The formula "the national economy does not need profit, but products" had been universally applied for many years. That formula fixed in people's minds the notion that the cost component of economic activities is unimportant. A failure to fulfil production quotas directly reflected on the work of industry and caused difficulties in the national economy units connected with them. If more resources are used than have been assigned by directive, the interests of other enterprises may only suffer slight adverse effects, and the damage is not locally serious but is inflicted on the national economy.

A major application of the motivation mechanism is to be found in the stimulation of the labor contribution. Its role increases when we go down the hierarchy levels where the stimuli connected with retaining one's post become weak or do not operate at all.

Formally, the system of remuneration for labor is strictly regulated. But the processes of allocation among enterprises of skilled and unskilled labor qualifications can be managed only to a limited degree. The centrally established rates of pay and patterns of supply of labor do not correspond to the real demands of enterprises.

An increase in the work force is the simplest way to increase output. Emphasis on stricter sanctions for failure to fulfil quotas, rather than provision of additional resources, prompted a serious shortage of labor as early as the 1970s, even through the demographic situation was favorable. Enterprises made greater demands for labor than could be supplied. For instance, in 1981–1985 the number of job vacancies left unfilled for lack of qualified applicants had increased considerably. While earlier only unskilled manual jobs remained vacant, nowadays the shortages of skilled workers such as machine operators, miners, metal workers and drivers have got worse.

Under such conditions it is inevitable that enterprises should compete for labor resources. Offering wages above official rate happens when the real value of labor is different from the value officially defined. The remuneration of some trades and professions is easier for the state to control, than that of others. Remuneration for piece-work is the most difficult to control, and in this, enterprises have considerable freedom of manoeuvre. And it is only natural that the remuneration for that group of workers increases at the highest rates. The disparity in wages of piece-workers and of time-workers makes enterprises put up the wages of the latter group with bonus schemes. The salaries of engineering and technical staff, and of workers in the non-productive sectors of the economy are the most rigidly controlled: while the remuneration of piece-workers grows automatically, in their case it is necessary to find new resources to restore their pay to the previous level.

The correlation is also changing between rate of pay and real labor contribution, even for those workers whose wages are strictly controlled. The fact is that it is very difficult to control the real intensity of labor for these workers. The proportionality between the wage scale appropriate for their contribution to the economy, and the established guidelines on wage differentials for different trades and professions is re-established through the relatively low labor intensity of their main jobs and through the development of various uncontrolled forms of individual labor activities.

There are only limited opportunities for regulating the dynamics of the national wages fund. The insufficient influx of labor to the primary industries has led to extraordinary measures aimed at increasing wages here. That, in its turn, aggravates difficulties in other parts of the national economy. Thus, an increase in the wages of miners at the beginning of the 1980s immediately caused a deficit in labor in the other industries of the coal regions including those directly servicing coal-mining itself. Here too rates of pay went up.

The pressure to maintain income bears directly on the effectiveness of inducements to work.

Imbalance in the consumer market forces industries making goods that are in short supply to compete to attract the labor force. One of the major factors determining the allocation of labor to enterprises and branches of industry is the priority placed on the goods they make. Regional differences in product availability influence to a significant degree the redistribution of the labor force. And while in some cases the supply of the priority goods is by cash stimulation of the labor force, in others it is achieved by prestige considerations, which directly contradicts the aims of socio-economic policy and stimulates an undesirable migration (for instance, to capital cities).

Public funds are also becoming actively involved in departmental haggling. The terms of access to such funds vary considerably among various branches of industry and among enterprises. As a result, the effectiveness of the system of social guarantees is weakening.

A potential advantage of socialism is that it can widely use common economic motives, such as the identification of personal interests with those of the community, to achieve the subordination of one's own activities to the tasks of achieving social aims. But widespread corruption, and various well-known but formally concealed privileges are now found in society and have weakened this advantage. It would be an unjustified over-simplification to expect that any of the real changes here, which are so necessary at this stage of our development, could be ensured by administrative instructions.

A number of factors operating in the framework of the existing system of management are what really limits the applicability of common economic motives. The identification of collective and personal interests with social ones presupposes the potential improvement in the efficiency and performance of the production unit can in fact be made. But we are well aware of the difficulties the enterprises identifying such potential are encountering.

The contradictions between social and collective interests lead to the situation where workers who are actively struggling for social interests turn out to be "inconvenient" and do not get support in their own collectives. The possibilities of real involvement of the working people into production management are limited by the formal regimentation of the enterprises' activities, by the hypertrophy of bureaucratic procedures in distributing resources.

The most important factor determining the effectiveness of centralized management is the availability of information on real processes taking place in society, together with the ability to process such information and formulate the tasks for lower echelons of the hierarchy in an adequate way.

Among the indices used the most universal ones are those of cost or value. They make it possible to compare all types of income and expenditure. The well-known problem in the framework of traditional forms of management of the economy of correlation between cost and natural indices does not affect the enterprises' independence, or the use of commodity and money relations. Rather, it is a matter for the information handling ability of the economic center.

A well-known shortcoming of cost indices is that they can be raised artificially through the use of price factors, assortment shifts and, when using gross output indices, through "repeated counting" of material costs. It is not by chance that the fulfilment of plans as computed by cost indices (without taking into account subsequent corrective amendments) was, as a rule, more complete than when measured by natural indices. The effectiveness of using natural indices in macroeconomic management is limited by the fact that the most universal natural parameters express the consumer qualities of the product in an extremely inadequate way (for instance, metallurgical equipment might be measured in tonnes). As for attempts to use various relative natural parameters which reflect real consumer qualities, these lose their most important advantages – namely clarity and controllability, and entail the formation of a system of parameters parallel to the prices which opens up further opportunities for overstating the production.

The aggregated indices do not themselves make it possible adequately to assess the state of affairs in the economy. But, nevertheless, as soon as the implications of the growth rates, as measured by the major aggregated parameters, become known to lower economic subjects, they become immediately connected with the mechanism of power motivation and those parameters are inevitably reassessed.

The failure to fulfil the plan when it is expressed in aggregated parameters does not at all mean that national economic interests have been damaged, but if we do say this, it will undermine the efficiency of the mechanism of power motivation. Plan targets have been corrected in this very direction due to various reasons (including real ones). Imposing stricter sanctions for failure to fulfil plan targets, irrespective of the reasons, aggravates the tendency to make a fetish of aggregated parameters. Favorable reporting dynamics of aggregated indices has become a most important independent result of economic activities, the significance of which is not reduced by the real trouble in the corresponding sector of the economy. The hypertrophy of the role of aggregated parameters due to their convenience is becoming stronger in parallel with the increase in the number of the hierarchy levels which communicate information from the enterprise to the economic center and in the opposite direction.

In production units there constantly emerge such situations when a demand to maximize aggregated parameters comes into an obvious contradiction with definite social interests (what might arise is a proposal to reduce production of a commodity which is estimated in tonnes, perhaps by increasing production of cheaper and more economical equipment which however results in the reduction of the volume of the output in roubles, and the like). It is a significant factor sharpening conflict between common economic rules and power motivation.

A number of basic characteristics of the current economic system make the promotion of scientific and technological progress more complicated. There is, for instance, a connection between how people are motivated and today's economic results, consumers play only a limited role in the economic process and this makes it possible to produce outdated goods, while preventing accumulation of reserves and inhibiting product development. The main way to counter these characteristics is to create, as a priority, conditions for scientific and technological progress. That, however, makes the scientific and technological progress rather isolated, and would cause serious losses as a result of poor connections between the areas where science is strongest, and the economy as a whole. The monopolistic position of many scientific and technological institutions would become stronger too.

Compensatory redistribution of resources is applied to favor those branches of the economy whose high priority is accompanied by a complex of factors limiting the results and resulting in low investment returns. This contradicts the natural logic of the theory of effectiveness and leads not to a cut in resource allocation, but to an increase, to the detriment of those branches of the economy where investment brings better results. When the unremunerative industry is large in scale (e.g., agriculture), such a policy makes for considerably lower efficiency in using investment resources. Fulfilment of cost targets does not necessarily guarantee fulfilment of planned results in such a case.

At the macro level there is a conflict between the branch ministries, which put pressure on the big collectives for additional revenue, and the general economic management which seeks to balance budgets solely on the basis of efficiency. This causes chronic overstrain of the national economy, and the problem of the balance of capital investment is one of the major causes for reduced efficiency of centralized management.

The allocation of resources to enterprises and the demands made by other bodies are determined by haggling with higher management. These demands may not coincide fully with the approved plan, but the relations of one enterprise with others are formed on this basis.

The distribution of resources through the hierarchy causes a lot of serious regional problems, and places limits on the interaction of enterprises located in the same region. Regional needs are not met, notably, the need to develop a production and social infrastructure. At the same time the regional economic authorities possess effective authority to act for the managers of local enterprises, and are thus capable of redistributing and directing some of the resources according to local needs.

The position of economic managers is complicated further by the contradictory demands of the departmental and regional leadership, who may redirect resources towards meeting regional needs with inevitable worsening of production results. The best protection against such complications is for an enterprise to have reserves making it possible for them to meet the regional needs without damaging their basic production. That is an additional stimulus to maintain an extra number of personnel.

Having the right to resources does not necessarily mean that they will be received in practice. The orders for materials (quotas, warrants and the like) issued by higher authority and supply managers may be considerably larger than the supplier can produce, or may be prepared to make available.

In such circumstances the supplier has some opportunity to choose the most profitable and convenient customer. The behavior of the customer may influence this choice; readiness to amend dates and terms of supplies on the supplier's request, tolerance of failure to fulfil contractual commitments, ability to render the supplier various services or to redistribute some of the resources allocated to the consumer in favor of the supplier; all these will make a customer more attractive.

Thus, the economic mechanism differs greatly in practice from the one which has been fixed by law. Under such conditions economic managers work out a set of strategies whose extremes are "inert" managers trying loyally to abide by the approved law and having no chance of achieving the targets set by higher authority, and "businesslike, strong managers" who develop a system of informal connections, who are ready to break the regulations perhaps damaging other elements of the national economy but simultaneously gaining high results for their own collectives. The choice of strategy is influenced by the balance between the sanctions for failure to fulfil targets, and for breaking the law. An attempt to make both types of sanction stricter leads to a greater turnover of managers.

The effectiveness of centralized management of the economy is only loosely connected with the number of the directives regulating economic activities. But it does not at all follow from this that the importance of uncontrolled processes, the degree of the deviation of real economic life from the priorities of the economic center is fixed. The efficiency of the centralized management might be higher if the existing traditional social and cultural aspirations ensure high efficiency of administration while the scale of the economy is relatively limited. This would make possible a strict limit on the number of management links between the economic center and the enterprises. The essential factors here are: validity of the economic policy being implemented; making the right choice of the directions of economic development; ability of the major investment projects, to counter departmental pressure; having the correct amount of economic data.

The stronger the disproportions in the economy, the stronger the demand for foodstuffs and goods, and formal centralized directives merely serve the spontaneous and poorly controlled distribution of deficit foodstuffs and goods. That is why the balance of the economy, that is the correspondence between the volume and structure of demand and real productive capacities is the most important parameter determining the effectiveness of centralized planning.

The problems facing the Soviet economy are connected not so much with excessive centralization which to a large extent is of a purely formal nature, as with a growing deficiency of the traditional methods under the changed conditions. Excessive administration of the economic links is only one of the forms in which this deficiency manifests itself. Claims that centralized management of the economy works against the economic independence of enterprises are groundless. The socialist economy cannot function normally without an effective centralized management, without an extended economic independence of enterprises, or without real incentives for labor collectives to produce more effectively and to develop real market relations. Radical economic reform aims to solve all these interconnected problems.

2. Radical Improvement of Centralized Economic Management

Radical economic reform is a complicated, contradictory process but an objective necessary because there is no alternative solution for the very important economic problems facing the country. The ways development of the economy has been managed, in the USSR and in foreign socialist countries, have more than once undergone the essential changes that characterize an economic reform.

Long ago it was noted that when the development of the economy demanded that control over businesses be tightened, and that quotas and assignments of work should be used more extensively, system management automatically developed in that direction, sometimes contradicting the proclaimed purposes of the economic policy. When it was necessary to increase the flexibility of management, and to make businesses more independent, for the restructuring to be real, a thoroughly thought-out reform with consistent implementation of the new policy had to be effected. Unfortunately, it hasn't been done in full measure so far.

Proclaiming that economic reform is to be one of the purposes of economic policy does not guarantee its success. The basic tasks set by the September

(1965) Plenary Meeting of the CPSU Central Committee have not been put into practice, because of inertia, and resistance from certain social groups with an interest in the existing methods of economic management.

Radical transformation of the forms and methods of economic management, with restructuring of all interconnected elements of the economic mechanism bring with them a risk of losing control over economic development, and of the emergence of spontaneous uncontrolled processes which would undermine the reform. These risks are aggravated by the impossibility of an accurate forecast of the economy's response to the changes in management, and by the imperfection of the new management procedures. This explains the attractiveness of compromise, and of a "careful" approach to restructuring. The general pattern of the current economic mechanism is to be kept; the management of production and distribution by central directives, though these directives will be reduced in number, while excessive and unnecessary indices of production are eliminated. Simultaneously the system is to be supplemented by a set of economic levers providing incentives for businesses to accumulate reserves and undergo intensive development. Despite the existence of certain contradictions the economic mechanism is an integral and internally coordinated system. It is resistant to innovations which cannot be incorporated into the general logic of its functioning even though they do not envisage its fundamental restructuring. This to a great extent explains the slow and contradictory course of reform.

The practice of setting quotas to businesses by directive is an integral part of the management system and is intimately connected to other important practices. Even after the switch-over to judging results on the basis of economic laws of supply and demand, planning is still risky; the fact of having declared a profit can lead to the next assignment of work being less generously costed. The contradictions between the practice of assigning work and the free operation of economic laws manifest themselves as an accumulation of funds not connected with the quality and the effectiveness of the collectives' labor, and in a disparity between the true value of the work, and that defined by cost indices, etc. That is why the number of directive indices soon starts to increase while the stability of the laws is upset.

One can break this vicious circle and bring the economic mechanism into line with what is required only by comprehensive restructuring of its basic elements, that is, on the basis of a radical economic reform.

As determined by the June (1987) Plenary Meeting of the CPSU Central Committee, the basic directions of such a restructuring include the following:
– marked expansion of the independence of businesses, their switch-over to complete cost-accounting and self-financing, more complete responsibility for final results, the fulfilment of their commitments to customers, the establishment

of a direct dependence between a collective's revenues and the quality of its work, a general development of collective contracts in labor relations;

– a fundamental restructuring and improvement of the centralized management of the economy, concentration on strategic planning, the coordinated and balanced development of the national economy as a whole and, simultaneously, resolute elimination of interference by the center in the day-to-day activities of businesses;

– a fundamental reform in financial planning and price fixing, a switch-over to full scale trade in commodities and the necessities of production and a restructuring of the management science, technology, foreign economic relations, and labor and social processes;

– a democratization of the present excessively centralized system with the development of self-management, and new emphasis on personal development and achievement with a clear delimitation of the functions, and a fundamental change in style, of party, Soviet, public and economic organizations.

The most difficult problem is how to move the new conditions under the current of affairs in the Soviet economy, which is seriously unbalanced.

Under current conditions there are many factors which cause supply to fail to meet demand. But it would be incorrect to conclude that in the framework of this mechanism it would be impossible to achieve balance. In practice in the socialist economy there are not merely two conditions – of balance and imbalance – but a wide spectrum of degrees of equilibrium with investment activities, rates of production and consumption and the satisfaction of the consumer goods market each deviating from balance for a different reason and in a different way.

Under the administrative system of management it is really difficult to ensure a flexible response of the production to the change in demand, to meet excessive demands for goods such as arise when customers hoard, etc. But imbalance in investment fluctuates considerably under the influence of factors which are not directly connected with changes in the economic mechanism. Balance of supply and demand regarding consumer purchases depends on how retail prices and incomes are related, under the monetary policy.

State policy as applied to sectors of the economy where it can bring quick (though as experience has shown short-term) results is of the greatest importance for achieving balance. This policy includes, in particular, a centralized programme of investing in construction projects that are nearly complete, reducing the amount of construction, and suspending construction projects whose completion is not absolutely necessary to ensure the intensification of the economy. Such aims have been proclaimed more than once. The priority of that programme is of fundamental importance, with the readiness of the economic center to give up other objectives if they conflict with the goal of balance of investments. This is also the basis for the financial normalization of the economy.

The most serious problems are connected with balancing production of various kinds. It is difficult to control the flow of money as businesses can always borrow extra money, and simultaneously the possibility is limited of direct control over money for investment.

The situation can be changed for the better by combining an improved balance in the relative attractiveness to the population of consumer spending and of saving and investment, with stricter controls on credit. The positive influence could be exerted by restructuring those economic mechanisms which most increase the deficit of supply of goods. They include the use of gross indices which directly encourage enterprises to increase consumption of materials, or indices isolated from the ways in which planning, assessing activities and stimulating the economy actually take place. They also include planning distribution of goods to be produced by new factories, general distribution of products that are not in short supply, and the like, and correcting deficits arising from imperfectly placed orders, etc.

It is impossible and unattainable in principle to balance the economy while the switch-over to new methods of management is in progress. Our theme is the necessity of limiting imbalances, and of creating the most favorable environment for restructuring.

In the current system of management, money without power has lost its ability to play the role of a universal medium of exchange. Turning it into productive resources is possible only with the permission of a higher authority and/or when other benefits and services are involved in the exchange.

A marked increase in the purchasing power of the rouble, and the elimination of the deficit constitute the backbone of the forthcoming reform. It is imperative to reorient the present-day management from its business of distributing material resources, towards managing financial and credit resources, and towards controlling the total demand in parallel with extending the rights of businesses to control their own finances.

The economic literature often ascribes the key role in the solution of that problem to comprehensive financial plans covering all flow of money and credit. A greater emphasis on achieving balance in the economy, and restriction of investment activities should all feature in the five-year plans. However, we believe that the importance of another planning document shouldn't be overestimated.

Experience suggests that planning availability of credit might destabilize the economy. The amount of credit appropriate for maintaining balance is difficult to determine, even for a short time. If overestimated, unjustifiably large amounts of money are attracted into investments due to the growth of the money supply, with increase of inflation.

The main task here is to place on those offering credit, responsibility for ensuring stability of monetary turnover, for maintaining the purchasing power of the rouble. Further, their role in making the very important decisions about the state of balance must be emphasized, and the automatic granting of credit to businesses out of the state budget must be blocked.

Incentives to make businesses more efficient in the use of resources, giving up petty supervision of their activities as determined by the mass of departmental regulations, all reduce the scope for using authoritarianism in organizing the interaction of businesses with the economic center. The strictest sanctions must be applied if financial results are unsatisfactory.

The experience of the switch-over to the New Economic Policy (NEP) proved that when the trusts transferred to cost-accounting and were relieved of state support and control of supplies, any inability to pay suppliers or the work force threatened stoppage of their activities, and this changed immediately their attitude to the administrative directives coming from higher authority. The problem of how to guarantee the independence of a business, so urgent for us at present, was to a great extent solved automatically.

Of course, one must take into account the restrictions imposed on the economic freedom of enterprises by social considerations, and by the overall balance of the national economy. However, it does not follow that we can have no strict sanctions for enterprises which use resources ineffectively. Extending economic independence to businesses whose freedom is very limited would cause serious disproportions in the economy.

Unprofitable businesses might have a function in the national economy, their losses being covered by guaranteed, long-term state subsidies. But even these do not have an unlimited and uncontrolled right to public funds.

If an enterprise cannot meet its debts even when in receipt of state subsidies, it is clear that there is a discrepancy between the quality of its work and the requirement of the community. That is why an inability to continue trading without external funding can justify extraordinary measures against that enterprise: credit, financial and governmental organizations must investigate and take decisions on the future of the enterprise and its property, and on the desirability and means of paying off its debts.

Enterprises may have many forms of economic responsibility. If trading conditions are unfavorable, insolvency may be a consequence for a business but this need not be the same process as the purely market mechanism of the automatic liquidation of bankrupt enterprises.

We have an opportunity to regulate this process in a planned way, and insolvency need not necessarily lead to liquidation at all. Bankruptcy might be averted by a change in the enterprise's specialization and trading practices, a reduction in output volume and a reallocation of resources to its most effective departments.

If there is centralized regulation of the redistribution of resources, should there have been bad company management, decisions on reallocation of resources could consider the whole range of socio-economic consequences. If a system exists whereby insolvent enterprises could be liable to restructuring, their attitude towards financial activities will change, with dramatic increase in the effectiveness of economic levers. It also opens up wide opportunities for using other forms of economic responsibility.

Banks have powerful influence on the activities of businesses and their policies can exert considerable influence on the development of the national economy as a whole. Extending to businesses the right to use their own money will free banks from minor activities such as handling financial transactions of companies. They will have to adopt a flexible credit policy, using credit limits and interest rates in order to maintain the balance within permissible limits. Such limits should be defined by analysis of the economy, reference to holdings of stocks in various businesses and of various kinds, and by such indicators as the level of debt and the rates of price increase.

All this will help to re-establish money as a universal equivalent which can be easily converted into other resources and products.

Methods of business management, in particular those relating to supply and marketing considerably in this situation. If raw materials, semi-finished goods and suchlike can be bought as required, this will eliminate the need to hold excessive reserves and funds. At the same time, a delay in the completion of the order, or its supply to a consumer who turns out to be insolvent will tell directly on the cost-accounting interests of the enterprise. If reform succeeds, the hero of the economy based on deficit – "a provider" capable of turning money which costs nothing into real resources – will inevitably be replaced by a commercial traveller whose main task is to overcome obstacles blocking the turning of the product of the business into the universal equivalent, namely real money. Wholesale trade will acquire a most important role, and certain measures are today being taken for its parallel development. However, the planned spread of that mechanism throughout the national economy requires fundamentally new approaches. For the time being, it is supposed that businesses will place orders with territorial agencies of supply and service and receive corresponding commodities from them. There are no grounds to believe that in the national economy the sum total of all such orders (even provided the macro balance is ensured) will correspond to the structure of the production which is determined by the economic interests and the productive capacities of the suppliers.

This contradiction can be solved following the logic of the correspondence between the interests of the producers and of the consumers by creating a framework in which enterprises independently form production and economic links, observing whatever restrictions are imposed by the state, such as choice of product type, preferential supply of priority consumers, etc.

This requires the monopoly position of supply organizations to be overcome, with basic restructuring of the established system of supply and service.

The task is to create an effective territorial and specialized wholesale trade servicing the national economy with direct connections among enterprises, supply and sales through cooperatives formed by enterprises, etc. under conditions of competition.

Further, a commodity exchange is an effective mechanism of regulating production and purchases, of providing up to date information on demand and supply, and of controlling prices and limiting monopolies.

Wholesale trade cannot function without commodity exchanges or the like, to coordinate demand and supply. They have an especially important role in ensuring priority supply of the most important consumers. Commodity exchanges will help overcome the limitations of interregional trade in consumer goods, such as the tendencies to "regionalize" the market which are possible under the reform.

A complete restructuring of the system of pricing is basic for coordinating the interest of both producers and consumers of goods. The drawback of the existing system of wholesale prices is that they are based on the cost of production, with a poor reflection of the real national economic effectiveness of the product, inflexibility and lack of relationship to the balance of demand and supply in the economy. Now the new general principles of pricing have been formulated: prices must reflect the effectiveness and the consumer appeal of the product, with fixed, discount and contract prices depending on the type of the product.

Only if prices are orientated towards balancing demand and supply will there be a flexible coordination of contradictory interests of businesses. Only a few very basic products whose prices are centrally controlled will be of this type, and the overwhelming majority of goods will be priced directly by their producers and consumers. Of course, we do not speak of a process that passively follows the market. Centralized price formation for the most important products, determining the permissible range of price fluctuations, calculating permitted prices with progressive taxation of excessive profits, are natural means of market regulation. They must be supplemented by the establishment and control of reserve stocks of the major products, with intervention buying to limit price fluctuations on the market, to prevent deficits and surpluses by stimulating production of the goods that are in deficit.

Economic independence of businesses, flexible coordination of day-to-day activities, prices aimed to equalize demand and supply are often considered to put stronger inflationary pressure on the national economy. The growth of prices may be attributed to inflationary pressure but may, in fact, be caused by such factors, such market-governed processes as more difficult conditions for mining increasing the prices for basic resources, improved quality and greater consumer appeal of the product, and broadening the range of goods on offer in response to new demand. Within such limits a planned, controlled increase in prices might promote desirable structural changes in the national economy.

Inflation proper arises from the existence of demand which has no supply. Inflation, as has already been noted, does not directly arise from the economic independence of enterprises even when they are free to fix their own prices. Such independence is likely only to influence the superficialities of the inflationary process.

Under reform inflation will not cause budget deficits as is the case at present, but increases in prices. High rates of inflation have a bad effect on people with fixed incomes, and the ability of increased production to deal with inflation depends on the effectiveness of control over demand.

That limits the choice of the investment policy to be pursued in the period of the economic reform. To use the resources released by reform to increase the volume of capital investments may lead inflation beyond the limits which can be dealt with by the economic regulators. To prevent this, it is necessary to pursue a strict credit policy, to make a wide range of chronically insolvent enterprises healthier, promptly to redistribute reserves in favor of such businesses as can use them effectively. This way might be feasible, but only if it enjoys great public support.

The idea that having a balanced economy eliminates the necessity of the direct management of production and distribution of material resources is oversimplified. The use of compulsory orders and quotas might be reasonable when regulating the production of goods of special public significance, while managing monopolistic businesses in such cases when an increase in prices for such goods to the point where there is a balance between demand and supply is unacceptable.

A distinctive feature of the system which is to be added during reform is that orders and quotas stop regulation all the business interconnections and turn into an auxiliary lever of management to be used for carrying out such tasks whose solution by decree is not effective. That is why such orders and quotas should be in such forms as do not infringe upon the cost-accounting of enterprises, and they should envisage the responsibility of higher authorities for providing resources, to meet the planned demand. The question is not whether five of fifteen directive indices will be given to enterprises, but that directive assignments of a traditional form, assigning a role responsibility for businesses to the center, must not be used.

State orders as a form of interrelation of higher state authorities with enterprises and businesses, are adequate to the new conditions. They can regulate the relations of the parties, in particular, by ensuring the supplies of resources necessary for the manufacture of the products ordered, for sales, for covering expenses and ensuring for the business a suitable profit, etc. If these requirements are fulfilled, state orders will often be among the most profitable, and businesses will compete to win such orders.

A serious threat to reform would be the direct regulation of economic activities, and state orders could become in fact traditional directive addressed assignments, in disguise. This really happened in 1988.

There still exists, as before, planning by directive of the production and distribution of even such goods the demand for which is fully satisfied. National economic balances are developed into a system of state orders for ministries and then for businesses. This procedure involves thousands of offices and hundreds of thousands of people. A considerable part of the management personnel does not even conceive of any other forms of economic management.

That is why it is necessary to place strict upper limits on the share of production to be covered by direct regulation both within the economic branch and at separate businesses. Purchase of products by the state over that limit must be on the basis of equal contractual relations.

At the first stage of reform the system of financial advances should be preserved only when some types of product are concerned. The most effective means to restrain its expansion is to pursue a stricter credit policy when such tendencies appear. Besides, it is expedient to distribute some of the goods financed this way through the commodity exchange at equilibrium prices in order to ensure enterprises permanent access to the necessary resources.

Wholesalers, whose job is to satisfy priority public needs, by distributing products in short supply should be controlled by direct methods in order to minimize abuses of their favored position compared with that of the retailers.

An important prerequisite for the normal functioning of the connections between economic units is that an active role should be played by the consumer. Here, of primary importance are the market structure, and the deforming influence of monopolies on it.

There is no doubt that the concentration of production is one of the general principles regulating the development of productive forces, and is a factor limiting the effectiveness of the market. But one must be realistic when assessing the scale of this concentration, as very many factors could constrain the functioning of small and medium-size businesses in the economy.

Large scale production does not always ensure economic effectiveness, with some notorious examples in this country. Large state organizations supported by a very rich resource base, whose monopoly has been secured by administrative measures, often prove to be incapable of competing successfully with individual and small cooperative producers who function only partly legally. This is often because of the large organizations being inert and having only poor control over real processes taking place in the lower parts of their structure.

We believe that a clear distinction should be made among the factors determining the monopolistic position of economic organizations.

In some cases major economic systems arose for organizational and management reasons: many of today's branch ministries and the intermediate links reflect the logic of a management system based on detailed regulation of the businesses. Such monopolies are strengthened further by the regionalization of markets, and by the administrative regulation of the goods among regions. Enterprises have only, as yet, limited opportunity to satisfy their needs through participating in the international division of labor which exists within the CMEA countries' market, and further afield.

Restructuring of the economic mechanism could and should be accompanied by the reorganization of such "administrative monopolies", which could be accomplished through the abolition of branch ministries.

It is clear that branch ministries have promoted strong monopolistic effects in production, suppression of the interest of consumers, increase of inflation in the national economy, and the priority of production and economic criteria over community and environmental interests. The branch ministries have become due to their socio-economic power, the main source of departmental interests in the economy. One cannot disregard the logic of the development of socio-economic processes. Ministries, as the experience of the reform in the 1960s showed, will do their best to regain all the rights they have lost, and to regain command of businesses and control over resources and their distribution. In our opinion, many branch ministries are a real hindrance to the extension of the rights of businesses, which is why they must be abolished and only ministries of general function and ministries managing the infrastructure and major economic sectors should be preserved. The solution of this problem requires the Law on State Enterprise to be changed immediately.

In cases where the formation of major economic complexes reflects progressive tendencies in the development of production, that is, it increases effectiveness and accelerates scientific and technological progress, this monopolistic position has an effect on the methods of management. The existence of such "monopolies" should be considered as conformable with the general differentiation of economic activities and, correspondingly, with the differentiation of the ways enterprises function in various parts of the national economy.

So far, only the first steps towards a closer connection between the internal and the external markets have been made. An increase in the purchasing power of the rouble, and its internal convertibility will create prerequisites for its external convertibility. Under the conditions of a new economic mechanism an enterprise should have the right to change roubles for convertible currency and, correspondingly, to change convertible currency for roubles according to exchange rates reflecting the foreign economic policy of the state. The integration of the country's economy into the system of world trade is, potentially, the most effective anti-monopolistic measure policy.

While working out the concept of the reform, it is necessary to take into account the real structure and scale of our economy. The specific features of the development of economic methods of management in our country will be determined by the following:

– the importance of the central control of basic industries, and of the defence industry because of the nature of these industries, where there are only limited opportunities for decentralization;

– the relatively closed nature of the economy and the necessity for major reallocation of resources from region to region and from industry to industry;

– the large scale of economic activities which greatly impeded centralized management of the numerous production units and which has placed restrictions on the use of market mechanisms, which are not a problem in foreign socialist countries. Those restrictions were caused by the following circumstances:

– the narrow market, the lack of real scope for organization of economic competition of enterprises in most sectors;

– regulation by processes which are not market mechanisms of industrial interconnections in the framework of the CMEA;

– insufficient attention being paid to economic factors closing down a business or even a separate production unit presents a considerable social and structural problem due to the scale of the economy.

A considerable extension of economic interdependence of enterprises in regulating their current activities is of fundamental importance for our economy. We have definite opportunities to ensure a dominant position of the consumer and a high economic responsibility of enterprises.

World experience confirms that an effective co-existence of industries is possible. The management system in any particular part of the national economy must be logical and an effective mechanism ensuring the interaction of various industries is of fundamental importance.

Businesses manufacturing products of special public significance will have certain special features, notably that a large part of the output will produced by state order and, there will only be limited opportunity to form a production programme, as there will be greater responsibility of state organizations for their activities.

Any extension of the rights of the industries which are natural monopolies, for instance, railways, electricity generation and communications, in selecting among their consumers, and in fixing prices may lead to an unjustified redistribution in their favor of a part of the income created by other industries. That is why conferring greater financial independence and abolishing unjustified restrictions in wage rates remuneration must be combined with effective control of prices and tariffs, with their activities being assessed not by gross indices but by the indices reflecting customer satisfaction, the observance of contracts, and the abolition of unjustified privileges of monopolistic industries as a way of compensating customers for breach of contract.

Releasing state organizations from routine management of enterprises enables them to concentrate on development, technological investigations, and on directing investment into science and technology. Success in these activities should be the basis for assessment of their managers, and of their ability to command the respect of the public.

As is known, at present management of science and technology is not effective. The bodies supervising the programmes do not have any real levers to influence the organizations involved in them, they are not actually in charge of the resources. To grant them such rights and resources would mean, in fact, creating a new layer in the economic hierarchy, and those are already quite abundant enough.

A balanced and regulated market opens up vast opportunities for using programmes to direct the development of production. That envisages the establishment of provisional agencies fully charged with the implementation of the programme in question, the allocation of fixed amounts of financial resources to them out of the All-Union, republican or local budgets. Such agencies should be granted the right to distribute credit, currency, and deficit material resources if needed, and they should be fully responsible for implementation of the programme. The programming agencies will be able to sign contracts with businesses on a competitive basis, which will state the terms of their participation in reaching the goals, in distributing the resources allocated to the programme as well as the terms and the procedure of their reimbursement.

Sometimes, economic reform under socialism is seen as having an intrinsic contradiction, in that the mechanisms of regulating current and investment activities are distinct, since centralized regulation of development limits the opportunities for coordinating demand and supply in the framework of the current turnover. We believe that there is a definite problem which can however be solved; it is the necessity of coordinating current and long-term tasks in the activities of businesses.

The adaptation of production to the structure and dynamics of consumer demand does not give businesses complete freedom of manoeuvre in the investment sphere. For that purpose it is sufficient to enable enterprises to make short-term investments aimed at adapting their productive capacities to the structure of demand. Businesses which utilize resources most effectively and best meet the needs of the customer should have a chance of developing faster.

This would happen if the means allocated to businesses depend directly on their profits, and on the amount of their amortization deductions; there are broad opportunities for using credit to finance the investments which could soon be repaid; the centralized investment funds allocated for the development of businesses are to be distributed on a competitive basis and, as a rule, they are to be repaid.

To state that because there is a connection between the coordination of the current economic activities and the regulation of investment flows, the market mechanism must be applied to the investment sphere as a whole, means presenting the situation extremely one-sidedly. The experience of the capitalist economy in past decades convincingly demonstrates the insufficiency of the market mechanism for regulating the processes of economic development.

It is obvious that the real market is not at all one based on perfect competition, and as that real plans have little in common with the ideal ones that exist only in abstract models of the socialist economy. That does not at all mean that we cannot use those positive features of the real market which we need so badly at present – the flexibility of the economic connections formed within its framework, the important role of the customer, the swift response to changes in demand in order to fulfil tasks of intensification. Also the shortcomings of the established forms of planning do not necessitate giving it up, provided of course it has been radically improved. The theory of optimum functioning of the socialist economy may become a theoretical and methodological paradigm for solving the problems of improving the economic system of socialism.

3. The Problems of a Switchover to Real Cost-Accounting

The success of economic reform will depend greatly on whether we will be able to use more efficiently production resources than has been possible under capitalism.

There have been numerous attempts, following traditional ways, to compensate for the lack of interest on the part of businesses in solving that problem by setting directive assignments aimed to improve the use of different resources and to achieve scientific and technological progress, as well as by attempting to coordinate the mechanism by forming incentive funds. The experience of the development of the economic mechanism in the USSR has convincingly proved that the introduction of more and more directive indices, which is the simplest response to the emerging economic problems, weakens the effectiveness of control over each index, and makes the system of incentives complicated and difficult to understand for the labor collectives, and contradictory by definition.

A fundamental restructuring of the motivation mechanism is needed, with directives no longer being used to stimulate effectiveness, and a switchover to complete cost-accounting being used to ensure real economic advantages to the collectives which utilize resources effectively.

For a long time the discussion of problems in the economic mechanism in general and particularly in the mechanism of stimulating effectiveness has been based on the implied premise that the established system of management functions satisfactorily and that it corresponds in principle to the socio-economic tasks facing the country. In this connection some authors have noted that the putting into effect of economic methods of management, with a switchover to complete cost-accounting will make it possible to raise efficiency and to utilize resources in a better way. Other authors have paid attention to the contradictions which inevitably arose while coordinating the wages and funds aimed at financing the development of production, with the final results of the activities of the enterprises. It was concluded that such a strategy of the economic mechanism's development was incompatible with the principles of socialist economic management.

That there exist, in our economy, considerable resources which can be used to enhance the effectiveness of the mechanism of material incentives, is not merely a hypothesis. It is a fact corroborated by the way our economic life is practiced. In businesses where material incentives for the effective use of resources are emphasized final results increase dramatically. But there can be no miracles in the economy, therefore the use of these resources in the national economy as a whole demands that a range of complicated problems should be solved first.

Nowadays, wide application of complete cost-accounting is not a matter of choice, it is an urgent demand of the time. But that is just what makes it necessary to pay special attention to the contradictions emerging during the period of the switchover to complete cost-accounting, and to acceptable ways to resolve them. A basic feature of complete cost-accounting is a clear coordination of the amount received by the enterprise for wages and for product development, with the results of its economic activities. That mechanism makes the volume and the structure of the enterprise's requirements depend directly on the receipts from production. But there are no grounds to believe that such a structure will coincide with that determined by the "bottle-neck" style of management, which uses directives to allocate scarce resources. Under such circumstances some businesses are forced to accumulate savings, or non-realized residues of cost-accounting funds, while others need more credit as a result of their own resources being insufficient to meet production plans. This contradiction can be settled provided the macrobalance of demand and supply is ensured and the mechanism of the regulated market functions, while regulation by directive is applied only to very important products that are in short supply. In such a case, the demand of a business for resources which are not available can be re-oriented to whatever resources are available. Otherwise, the effectiveness of the stimuli introduced is undermined and simultaneously the imbalance in the national economy grows.

In this situation serious disproportions also emerge in wages. If the profitability of a business is based on fixed prices which poorly reflect the true value of the product to the national economy, and on other external factors such as the availability of materials and the reliability of suppliers, the influence of wage rates on the quality and the effectiveness of labor is weakened.

A switch to complete cost-accounting will have stable positive results only if there is a fundamental restructuring of the mechanism, with day-to-day coordination of the activities of businesses.

The practice of permanent revision of quotas, targets and limits and their adjustment to supply and demand had nothing in common with complete cost-accounting. In fact, these traditional methods served as a form of limit distribution of finances and of removal from circulation of the resources surplus to the budget. That is why the wish to regulate the distribution of the gross income among the society, businesses and to various special funds by means of these parameters is quite understandable and justified, but we do not share the utopian belief that such parameters can be established individually for all elements of the economy once every five-year period, to remain stable thereafter. The dynamic nature of the economic processes actually needs changes in medium-term planned provisions. When only a few experimental businesses are concerned, it is easy to ensure the stability of the parameters at the expense of other parts of the economy. But if the regulation of financial flows becomes one of the most important levers in the centralized management of the economy, it is next to impossible to maintain the stability of the parameters in real life.

The stability and effectiveness of cost-accounting are best assured by not allowing the changes to appear to be directed towards individual businesses, apparently to justify requisition of the reserves of the enterprise. This can be achieved only through a switchover to the system of profit or gross income taxation of enterprises.

A switchover to complete cost-accounting implies regulating processes which have so far regulated by individual decisions through applying parametric methods. Let us consider the problems which arise in this connection regarding three uses to which the gross income is put, namely wages and incentives, funds for development of production, and payments to the state budget.

The interests of the collective of a socialist enterprise are closely connected with the preparation of a budget for wages, incentives and bonuses. If these funds are formed as a residual value, the difference between the total revenue of the business and the costs of production, including the budget for development, the connection between the effective utilization of resources and wages is closest. There is no need for encouragement of economy in materials, energy and equipment, or of encouraging mobility among trades; all these processes bear directly on earnings. The necessity to regulate "from above" no longer arises either. Dealing with overexpenditure, fines, accumulating extra stocks must be done at the expense of the wages fund.

At present the most important task is the struggle against uniformity of wages, and in favor of differentials in wages reflecting differences in labor contribution. But the basic problem is that wages change in ways which are extremely sensitive to social conditions.

Opinions regarding appropriate wages, for a particular job, permissible degrees of wage differentiation in general and within trades and professions change in the process of socio-economic development. Such altitudes can be altered, but at present they must be taken as given. The strength of these altitudes must not be underestimated, and attempts to improve differential wage rates may go beyond what is currently socially acceptable and this may lead to lower labor quality and poorer discipline among workers who would be disadvantaged under such a system. Consequences would include aggravation of social tension and inflation.

It is unconvincing to refer to the principle of distribution to individuals according to their work, as is often done to argue that a sharp growth of differentials in wages is possible and necessary. The correspondence between differentials in wages, and in the different values of the work done, is not quantitative, but relative.

In restructuring the system of allocation of the funds used for wages we must not only ensure a direct dependence on the results of economic activities (not difficult in itself), but simultaneously make sure differentials are within socially acceptable limits. It will be regulated to a greater and greater extent by a thoroughly thought over system of taxation.

The wages fund directly and obviously reflects the results of economic activities. There is no more effective mechanism to eliminate overexpenditure to arrange matters so that waste and overspending result in depressed wages while economy and efficiency elevate wages. However the residual fund out of which wages would be paid is one of the most volatile entries in the balance sheet.

There is no single, simple solution to the problem of relating wages to performance and taking into account economic factors beyond the control of the work force. Clearly, it is important to outline the range of permissible values, to establish the parameters of this or that wages formula which will depend on the emerging economic situation, and on the specific features of each enterprise.

The major factor basic to a discussion of the forming of a wages policy is the institutional system of business management. The greater the control over the business that is exerted by the labor collective and by its elected representatives, the more easy it is to ensure a strict dependence of labor remuneration on production results. Making the labor collective responsible for the results of the economic phenomena which it cannot influence directly is a potential reason for serious social conflict and does not make sense in economic terms.

We shall stress once again that in those businesses where there exist the greatest possibilities for the immediate participation of the working people in solving economic problems, where real responsibility can be given to labor collectives, the mechanism of payment of wages from the residual fund can be used. The excessive growth of individual earnings should be limited at the same time by a system of progressive taxation of any individual earnings which exceed a certain level. This would relieve the mechanism of wages calculation from the duty of ensuring social guarantees and regulating wage differentials.

At major enterprises manufacturing goods of high social significance the right of the labor collective to make important decisions determining economic activities would be limited. It is right and logical that in such businesses money should be allocated for wages in a way which is relatively independent of economic performance.

In our opinion, wages policies of many intermediate forms should be used, between these extremes. They would permit coordination of the wages fund with the profit and the use of normative regulators intrinsic to the mechanism of calculation these funds, which would restrict the range of differentials within socially permissible limits.

With a growing variety of forms of wage regulation, the redistribution of labor resources will be made easier.

The increase in the mobility or the labor force can be a very constructive factor. Attitudes towards remuneration and individual contribution differ greatly according to age groups, and different for men and women. [2] Nowadays, workers who would prefer to earn considerably increased wage through working harder, and who are not very concerned about the risk of a decrease in their earnings if their company does badly, have no chance of putting these preferences into effect in public production and have to consider self-employment. The formation of a wide range of business types in the public sector which work on the principle of close coordination of labor remuneration with results, will create favorable conditions for such workers. Any attempt to retain this kind of worker in businesses where there are strong social guarantees and a lower range in wages are not advisable. On the other hand, many workers do appreciate wages that are predictable and regulated.

Greater responsibility of enterprises for the financial results of their activities and their solvency will work to improve existing distribution of labor resources and reduce overmanning through a closer coordination of labor remuneration with the results.

The situation emerging at present in the field of labor resources, with the serious shortages in a number of the branches of the national economy which are difficult to overcome in the medium-term, makes fears of a possibility of mass unemployment quite abstract. In the long-term such a threat will be removed

---

[2] T.I. Zaslavskaya, The Creative Activity of the Masses: Social Reserves of Growth, The Economy and the Organization of Industrial Production, 1986, No. 3, pp. 3–26 (in Russian).

because the socialist state retains the means of managing structural policy and investment activities.

Socialism is not a charitable institution where full employment is automatically to be ensured through creating new jobs. At present it is necessary to move to a policy of effective full employment defining full employment not only socially, but also economically. The enterprises should be responsible only for the economic effectiveness of the workers in their work forces. State and regional authorities must be responsible for ensuring full employment as well as for the provision of employment to surplus workers. This all requires the creation of an effective system of refresher courses and retraining of the personnel as well as social security for the workers who are temporarily unemployed. This redistribution of functions must be accompanied by granting the corresponding state bodies the rights necessary to pursue an active policy in the field of filling job vacancies and creating new ones.

Extending the economic independence of enterprises into the sphere of product development, and combining their increased responsibility for the effectiveness of capital investments with more efficient ways of pursuing structural policy, are of fundamental importance for the achievement of complete cost-accounting.

In this connection some economists oppose restricting the financial responsibility of enterprises to simple decisions, and propose that even the creation of new productive capacities should be financed out of cost-accounted funds. Under this scheme, only the construction of new enterprises and infrastructure projects should be carried out using budget funds. Other economists believe the rights of enterprises should be confined to simple decisions, with a strict centralized control being exerted over investment for reconstruction and expansion.

We believe that a constructive solution of the problem requires a clear distinction of investment funds into those which are cost-accounted and which arise from the income of a business, and those which are at the disposal of the central organs of economic management.

If the gross income (profit) of enterprises is taxed and they have a chance of determining the uses to which their own cost-accounted investment funds are put, the issue of what proportion of the aggregate volume of investment resources of society is in fact derived from enterprises becomes fundamental. The universal nature of normatives does not allow local variation in the way social priorities are dealt with in the course of development. We mean the financial resources which are received and used by an enterprise are kept separate from the capital investments allocated by the structural policy even within the same enterprise.

Keeping the share of the resources the enterprise can control (the production development fund) lower than is necessary for ensuring maintenance gives rise to feelings of dependence, distorts the information reported to higher authority on the prospects of capital investments and relieves enterprises of responsibility for the results of their investment activities. The extremely limited rights of enterprises so far as investment is concerned impose strict limits on the prospects for extension of their freedom to make new production interconnections, and prevents flexible restructuring of production in accordance with changes in demand.

Making enterprises much more responsible for the cost-accounting and self-financing of the whole process of maintenance and growth on the basis of the normative distribution of profits, and allocating to them a major share of the funds to be invested in production (excluding resources for the construction of new enterprises and infrastructure) may lead to serious contradictions. Some businesses may accumulate considerable financial resources whose use is ineffective from the social point of view, while others badly need external finance for promising investments.

Of course, in the socialist economy there can also be a redistribution of an enterprise's resources for mutual benefit by creating joint ventures (including joint-stock companies) with special purpose loans for development. However, the prospects for basing such a scheme on the automatic redistribution of financial resources are restricted for the time being by the absence of the institutions that might enable the investor to control production and other economic activities including distribution of profits.

In the framework outlined, financing new connections between enterprises and the economic center is a task of a quantitative nature and it must take into account all factors relating to the direction of investment policy, the availability of capital to finance new construction, the expansion of the capacities of the enterprises, their reconstruction and retooling.

If an efficient enterprise accumulates financial resources which could be used to expand the production of goods in demand as defined by universal normatives applied to the distribution of the gross income, there are no grounds for limiting its investment activities. Even the construction of new enterprises must not remain a state monopoly, but can be carried out after a thorough study of its economic potential, perhaps by operating amalgamations or consortia, with joint-stock constitutions. Flexible credit mechanisms, "economic haggling", competition are all possible here, and there must be a choice of how to distribute expenses among all the participants including the state.

Any system of economic regulators may serve to conceal traditional hierarchic interrelations in economic life. Contradictions are inevitable between current methods of ensuring economic priorities such as giving directive assignments, regulating distribution of produce, granting financial and credit privileges, and the theory and practice of the reform. It is very necessary to use, carefully, the individual levers that are available under the system we have, and to maintain our general economic normatives. Reform is basic to socio-economic development, and its consistent implementation by methods which do not undermine the effectiveness of the normative mechanisms being introduced is necessary to do away with the emerging disproportions.

If higher authority independently assesses the activities, determines remuneration of an enterprise's managers, and makes decisions on their replacement, then no matter whether the priorities of that authority have been made known in the form of official directives or informally, they will a determine the policy of the administration. Authority's actions may be contrary to the cost-accounting interests of the collective while the extension of the rights of enterprises may be used by the management to achieve their own purposes. In that situation the normative levers used are addressed to one group of people while the real decisions are being made by other people who do not depend on those decisions.

The experience of foreign socialist countries has proved that abolition of all directive indices cannot guarantee the further economic independence of enterprises unless institutions are established so the management bears responsibility not only to higher authority, but also to its own collective, the leading trading partners, banks and other financial establishments. That is the only way to solve the problems of the limited horizon and responsibility of some managers. The interests of the labor collective as a whole, and of permanent trading partners are of a long-term nature.

Making the collectives more responsible for their economic activities may be assured by a cooperative form of organization, which is true of industry and of the services sphere, too. The restoration of the unjustifiedly liquidated cooperative sector here in combination with the legalization and development of socially controlled forms of self employment in manufacturing and services would relieve the state of particular, current problems, and enhance the balance of the market. To get small enterprises effectively involved in the system of public production is closely connected with granting them the status of cooperative enterprises, though major enterprises must be set free as well.

There cannot be universal solution of the problems of radical realization of socialist property. One must take into account the social importance and the scale of the production, the real opportunities for the control of the collective element over the administration. Otherwise, the oft-repeated stressing of the universal nature of the principle of workers' self-management may actually lead to its distortion, may permit for the emergence of uncontrolled bureaucracies and procedures using superficial forms of democratic discussions and decision-making as a cover.

Under the existing economic mechanism the administration's responsibility to the labor collective is low, and is compensated for, firstly, by the strict regulation of labor relations (the restrictions on rights to dismiss even obviously surplus, undisciplined or unqualified workers, the opportunity for the workers to appeal to higher authorities with the right to interfere in the enterprise, etc.), and, secondly, by lack of suitable labor resources. Practicing democracy in the work place creates the prerequisites for the extension of the rights of labor collectives to regulate labor relations. The collective is no less capable of solving these problems than higher authority or the law-enforcement agencies. Simultaneously, the social acceptability of the policy of economically and socially effective full employment is increasing.

Thus, the prerequisites for increasing the effectiveness of all the components of the motivation mechanism have been created. The prerequisites for power motivation are created through maintaining labor discipline, and through raising the prestige of the job. The prerequisites for raising the motivation to increase labor contribution are to be created through greater differentiation in wages to reflect the difference in the results of labor, through overcoming the imbalances in the provision of consumer goods, through ensuring effective employment of the population. Motivation arising from ownership is to be fostered by developing the democratic principles of management, and by overcoming the bureaucratic regulation of the economic activities. The most important thing is that it is difficult to expect any successful implementation of the economic reform without a thorough and comprehensive reform of the institution of socialist property. Without it we shall face only a deadlock and there cannot be any illusions about that.

4. Some Socio-Political Aspects of Radical Economic Reform

The central issue of radical economic reform, its purpose and the justification of all the measures taken, is a comprehensive increase in all forms of proprietary motivation of the participants in socialist production, through overcoming the existing and justified attitude to public property as a bureaucracy with strictly limited participation of the working people.

The development of workplace democracy, with active and effective participation of the working people in direct management promotes the solution of that problem. However, the isolation of enterprises from public production is, in general, limited. Enterprises carry out their economic activities in the framework of economic parameters formed by authorities which are beyond their control, while they themselves are inseparable from the socio-economic system. The development of workplace democracy cannot be opposed to the perfection of socialist democracy at the state, republican and local levels, to the development of the mechanisms of social control, and to a stronger political responsibility.

Full access to adequate information on socio-economic processes is an indispensable condition for the development of social control. Nowadays, the scope of the statistical information available to the public is unjustifiably limited.

In many cases restrictions and the departmental control over information simply camouflage economic miscalculations and the irrational use of public resources. The freely accessible statistical information is of a propaganda nature, demonstrating successes achieved rather than providing a real picture of the national economy to the joint proprietors of socialist property, which would reveal the existing difficulties, and the negative phenomena which must be brought to the earnest attention of the public.

So far it has been difficult even to get the information on how the results match with the targets of the five-year plan which had been adopted after nation-wide discussion. The disappearance of this or that parameter from the public statistics could serve as a reliable criterion of there having been trouble in that area and of a disparity between real processes and the planned targets.

Meanwhile, even in the most difficult years of Soviet power's rise and development the public and the specialists had at their disposal the information which made it possible to trace the real economic processes to the in the fullest detail, for instance, the most detailed information on the money supply, on the dynamics of prices, on the international exchange rates of Soviet currency (a ten-rouble gold coin) on the value of Soviet banknotes and gold coins, of gold and foreign currency on the black market in different regions, etc. The problems of financial and credit policy, and their interconnection with development were debated properly and constructively in the economic literature, and in the mass press. At that time the Party was not afraid that such information on the country's economy would play into the hands of our class enemies, and could be used in hostile propaganda.

At present the joint proprietors of socialist property have access to information on the number of amateur dramatic groups, on other clubs and on the distribution of their participants across the social spectrum, but they have no data on the revenue from the taxation of various types of production, on the money supply, on gold reserves, etc.

The summary report of the CPSU Central Committee to the 27th CPSU Congress noted that the dissemination of semi-truth and the side-stepping of urgent problems hinders development of realistic policies and interferes with our progress. The USSR Supreme Soviet, the Council of Ministers and the State Planning Committee do not possess trustworthy information.[3]

The most important task of centralized planning is to determine socio-economic priorities, and to distribute the limited public resources in various directions. The task is based simply on choices. The allocation of money to some public needs limits the possibility for fuller satisfaction of other needs. But at present that is poorly taken into account in the practice of management. Choices which are in essence of a socio-political nature and which affect the interests

---

[3] The Proceedings of the 27th CPSU Congress, p. 23.

of a wide range of the socialist society's citizens are often made on the basis of bureaucratic procedures in an attempt to create favorable grounds for the satisfaction of departmental interests.

The planning of socio-economic development is insufficiently orientated towards the most effective use of resources. Plans are worked out with only one, often ill-balanced set of options, which excludes any assessment and prevents choice of the best economic decision.

Another urgent task is basic reform of the state expertise. At present it is a part of our planning and economic organs and, as experience shows, it is often influenced by departmental interests and deals only with minor problems. In fact there is no expertise for planning economic and social development. The state expertise should be transferred to the jurisdiction of the USSR Supreme Soviet, the Supreme Soviets of the Union Republics and the local organs of Soviet power. Only on such a basis can the role of the Soviet authorities in the distribution and effective utilization of public means be enhanced.

Nowadays, all social problems have to be solved by expenditure of the proceeds of public production. Consequently when choosing among alternative uses of these proceeds, the most expedient ones should be chosen, and relapses into traditional thinking as manifested in propaganda cliches like "the state takes care of the growth of people's well being (the development of health care, the insurance of the stability of retail prices, etc.)" are absolutely unacceptable. Such reasoning is appropriate only in a description of the attitude of a considerate lord of a manor to his serfs, but it makes absolutely no sense when we speak of the socialist society. It is not the state that takes care of the solution of this or that problem, but the working people as joint proprietors must determine the most expedient ways for social and economic development, using the corresponding state institutions to work out and implement economic decisions. Only that model of interrelations is adequate to socialist democracy.

Introducing such an approach into practice, raising the efficiency of the democratic institutions of the socialist society, overcoming the attitude to socialist property as being purely a bureaucracy all greatly expand the room for maneuver in the social sphere. Such measures which an estranged, bureaucratic management could not afford to take become possible and acceptable when they are taken on the basis of a wide discussion and collective decision-making.

The situation in the consumer market causes special anxiety over the fate of the economic market; there exists an obvious tendency for the deficit in consumer goods to get worse. The tone of many letters sent to newspapers and the like indicates that if conditions for consumers get much worse there will be a real threat to restructuring. Lower standards of living, larger deficits are often associated in the public mind with the economic policy currently being pursued, namely economic reform, despite the fact that we are really dealing with the consequences of serious mistakes in the economic policy of the past decades.

Though, in all fairness, it should be stressed that in the past four years we have made lots of mistakes.

The attitude to cooperatives is characteristic in this connection. In 1988 state enterprises and collective farms supplied the trade with produce which cost 17 billion roubles less than envisaged by the plan. The total volume of the of the produce originating from cooperatives including cooperatives processing recycled resources came to 6 billion roubles. The turnover of state and cooperative trade was equal to 366.4 billion roubles and that of paid services was 61.8 billion roubles. As we can see, these figures simply cannot be compared in scale. But nevertheless, the idea that the cooperators bear responsibility for the worsening of the deficit in consumer goods has already become conventional wisdom.

Definite efforts are being made to improve the situation in the consumer goods market. Heavy industry including the defence industry are getting more and more involved in the production of consumer goods. But these measures have proved to be insufficient to cover the consequences of reducing the availability of alcohol and of imported goods.

The existence of a bad deficit on the market is not at all the best situation for the implementation of the economic reform. The lack of real competition makes it possible to charge excessive prices, to sell low quality products. It will take time for the positive changes in the national economy to influence the state of affairs in the consumer goods market. Meanwhile, people's confidence in the policy of restructuring, though still great, is not at all infinite. More than four years have passed since its beginning, people have been waiting for tangible changes in the standards of living which have failed to appear. Hence the longing of some people for "a firm hand" and the apathy and skepticism of others towards the possibility of real change. To keep society in favor of the fundamental socio-economic transformations which are currently underway it is necessary to overcome the negative tendencies in the consumer goods market in the near future. If the situation still further deteriorates, we shall face the following choice: either empty shop shelves, growing queues, rationing and a flourishing black market or an uncompensated rise of prices and, consequently, a lowered real income of the population. Each equally unacceptable, and what is more, each would be disastrous for restructuring and for the fate of our people.

The state of affairs in the retail trade is closely connected with the general financial situation in the national economy and, in the first place, with the state budget. The worsening deficit of consumer goods is combined with a sharp deficit of the state budget. The connection is obvious: excessive expenditure by the state causes an expansion of the population's incomes which are not covered by the stocks of goods.

Some untimely and for the time being economically non-feasible projects are financed out of the state budget, ineffective businesses and enterprises are supported alongside the normal expenditures stipulated by social needs. A greater budget deficit indicates the inability to redistribute the resources to those activities which meet the effectual demand.

At present attention is being drawn to the problem of subsidies for foodstuffs. Proposals to raise prices for meat and dairy products are often connected with the reduction of budget expenses. One must be well aware of the fact that only some of the problems connected with stabilizing the market can be solved in such a way while it is absolutely impossible to solve the problems of the budget deficit. The full-blown compensation necessary to avoid lower standards of living for the population will require a corresponding increase in some money payments to the population which are again to be financed out of the budget. The switching over of the demand from meat and dairy products to manufactured consumer goods, if it does occur and provided the price structure remains normal, will not improve the market situation in general.

As we cannot accept any lowering of the real incomes of the population or "economizing" in the social sphere, we have only two major items of budget expenditure which can be revised in principle. Firstly, military expenses and, secondly, expenses of capital investment and covering the losses of unprofitable enterprises. The prospects for economizing in the first direction are determined by a complex of foreign political factors which are independent of our economic activities. But their reduction as well as the reduction in the second group of expenses is now becoming an urgent demand. So far, the measures aimed at the financial improvement of the economy have been simply naive economically, to put it mildly.

For decades, attempts have been made to compensate for the faulty economic mechanism by directing a flow of resources to, for instance, the agro-industrial complex. The greater part of the corresponding expenses in whatever form had to be met by the budget and by credit which had lost its specific features and turned into a variety of inflationary budget financing. It is the state budget that pays for large-scale (but ineffective) water and land reclamation projects, for the construction of reinforced concrete "palaces" for cattle. It is from there that the funds for the rapid increase in supplies of low quality agricultural equipment came. The same source provided for the growth of production of mineral fertilizers and for covering the losses resulting from their misapplication.

There is no need to prove all this: only when a genuine master of land appears, when bureaucracies stop commanding agriculture, will the resources start bringing returns. At present it is important to meet the demand of agricultural enterprises only for those types of equipment and services that they are prepared to pay for.

State financing of large-scale investments aimed at increasing the production of raw materials, energy and fuels also needs basic revision. Many of them will start bringing returns only in the next century, but their serious burden on the budget, on the consumer goods market, on the whole structure of the economy is already felt now. Tactical and strategic aims coincide in this question: it is much more effective to achieve real changes in reducing the capital intensitivity of public production than to finance the plundering of natural resources. That is just what the centralized and decentralized economic structural and investment policy should be aimed.

The inventory of production projects currently under construction was started in 1988. The task is to achieve a real reduction of the construction period, to concentrate resources on the primary and the most effective projects. Such an attempt is made for the second time in the course of restructuring. The measures taken in this connection in 1986 brought only very modest results due to fierce resistance by the part of the branch ministries, the local party and Soviet authorities. Yet another defeat of the central government in this matter would have an extremely negative effect on the whole course of restructuring. The reduction of the construction periods is also necessary to lower the burden on the state budget considerably. It is especially important fully to provide with resources those construction projects which can ensure an increase in the supply of consumer goods in the near future.

Cutting down the purchases of imported consumer goods also inflicted a heavy blow on budget revenues as well as on the retail trade turnover. The growing physical volumes of our export of traditional goods (oil, petroleum products, gas, timber) as well as the changes in the export of engineering production are not sufficient to compensate for the worsening trade conditions. Though our external debt in convertible currency is still relatively small, its tendency to grow must certainly be stopped. At present the real situation in the national economy does not give us good reason to be sure that loans will be spent effectively enough so they can be repaid without problems. Another question is how to save currency. Significant import of capital and goods is possible only if we really carry out the economic reform.

It is difficult in the present-day situation to consider solving our foreign trade problems at the expense of public consumption. Withdrawing finance from the economically unjustified projects generated under the old mechanism of economic management, and in purchasing manufactured consumer goods to the annual value of 1.5–2 billion foreign exchange roubles could relieve the extreme tension on the consumer goods market, and would make it possible to reduce the budget deficit. Accelerated implementation of the economic reform in the industries supplying the consumer market could become easier: tough competition from imported goods would quickly reveal the enterprises which are utilizing the resources in a really effective way, which should then receive priority development. It is even more important to import more capital into the industries of the consumer sector, possibly by means of setting up joint ventures.

In the medium-term perspective the only sensible and clearly necessary direction for structural manoeuvre in our economy is towards the consumer: a big increase in investment in the production of cars and fundamentally new and technically sophisticated durable goods, together with development of service and maintenance facilities, in fruit and vegetable production on a massive scale, in the accelerated construction of glasshouses, and the expansion of direct sales of building materials, and all kinds of mechanical equipment to the population, etc. The analysis of the consumer goods market indicates this to be quite urgent. The sooner the people see the real impact of restructuring on their standard of living, the more irreversible will be the process of changes in our society.

It is also connected with the intensive development of the cooperative system of organizing production and consumption throughout the national economy and, of course, in the first place, with ensuring balance in the consumer sector. The development of the cooperative form of property ownership actually promotes a considerable extension of the economic independence of enterprises and amalgamations in the state sector of the economy, lowering the monopolistic effects by means of raising the importance of socialist competition. It goes without saying that all this requires a normal market for the products and a well-thought-out system of taxation to be developed. The extension of the right of the owners of socialist property, the real coordination of their economic and social interests, the determination of where they can be most effective are among the central problems of the purposeful implementation of radical economic reform, which is the basis of restructuring all the vital activities of our society which is at the turning stage of its political and socio-economic development. It would be immoral and politically irresponsible to call upon another generation to make sacrifices.

5. Outlook for the Development of Diverse Forms of Ownership in the USSR

The socialization of production and development of socialist ownership constitute a key aim of the economic policy pursued by a socialist state. They underlie the entire range of social measures aimed at improving the economic mechanism, accelerating socio-economic development, achieving continuous and firm growth in the national economy, and developing the class-based structure of society and the socialist way of life. A fundamental and all-embracing restructuring of social resources, proceeding from the need substantially to raise the efficiency of social production, makes it imperative for socialism to advance quickly to a new definition of the permissible forms of ownership, their scale and the rates of transitions to them. This advance, this leap forward, is designed to actively involve of all social groups and strata in economic, social, and cultural issues and to make socialist management more democratic by developing social self-administration. There is no other way to be found here.

The USSR has been developing for decades on the basis of a continuous socialization of production. Here are some of the resulting historical consequences.

1. A high level of socialization of the material and technical basis of production has been attained. Concentration, specialization and cooperation in production have led to the accumulated economic growth of industries and big enterprises and to a fundamentally new and more profound economic interaction. This interaction is not determined only by the fact that production structure and the entire system of economic ties among enterprises are growing increasingly more complicated. While exerting a long-term and increasingly more powerful influence on the socio-demographic and ecological factors common to them all (e.g. labor resources, environmental problems), individual branches of the national economy are also linked with each other through various processes which go beyond direct production activities and which do in fact constitute some major factors of socio-economic development.

2. A network of service industries has been established, which have "socialized" a considerable number of functions which were previously scattered among individual families. All-round "socialization" of these functions has affected the most sensitive aspects of social life (the family, the upbringing of children). More rapid growth of employment in the non-production sphere has resulted in a bigger part of the population working under entirely new conditions, with less regimentation than in the basic industries, and direct contacts with customers. Development of the non-production sector puts in the forefront such issues as diversity and flexibility of forms of socialization and the broad involvement of people not only as workers but also as consumers.

3. Essential material and technical and socio-economic prerequisites have been created for a transfer to a real socialist collectivization in agriculture, that is, the formation, in material and technical terms, of a single agro-industrial complex with a view to bringing closer together and subsequently merging the socio-economic structures of town and country.

4. The evolution of the Soviet economic mechanism has entered the stage where links among industries and regional management structures are evolving rapidly. The formation of organizational structures relating to socialization of production has been completed, which is indicated by the establishment of a network of production and research-cum-production associations. The allocation of resources to effective development of production based on new combinations of industries and on regional self-sufficiency is high on the agenda now. New forms of ownership are now involved in economic activity (industries, amalgamations, programmed management bodies, etc.). All of them consistently defend their own interests in decision-making, thus contributing

to the process of elaborating rational decisions which take due account of all economic, social and environmental factors.

5. Large-scale transformations in the social structure and the way of life have been effected. The growth of urbanization, mass involvement of women in socially-organized labor, higher cultural, educational and professional attainments of the larger part of the country's population have laid the groundwork for a qualitatively new stage in social and economic development. In social terms, this will bring about a really universal approach to man's working potential, providing for his active rather than purely passive involvement in life.

The change in the socialization of production, the basis for socialist ownership, has revealed a number of grave problems and contradictions. The majority of these problems are connected with the concentration of supervision and control over the distribution and use of resources in the hands of a system of sectoral management. The latter developed as a system of management relating to industrialization, and the accelerated solution of socio-economic issues at a time when the economy was in its extensive growth phase. Subsequent further development of industries has led, however, to a range of complicated issues and relations, which, under present conditions, stand in the way of intensified economic development.

Here we have in mind that kind of sectorial management which gave rise to the socio-economic phenomenon of "departmentalism" with regard to individual industries. This is a special kind of ownership that originated at the stage of extensive economic growth because the overwhelming part of production, as well as a significant part of the non-production resources, was controlled by various ministries and departments, while some other social institutions, which could also dispose of those resources, remained economically weak. There are three types of such institutions, the harmonious interaction of which constitutes a fundamental precondition for the effective functioning of the entire socio-economic system. Of these institutions (of industries, territories and goal-oriented programmes), only the first has practically a full monopoly of economic activity. Continued existence of this monopoly cannot be justified by any considerations related to production or scientific and technical efficiency, being fraught with dangerous social consequences.

The hopes that various ministries would carry out a common policy at the sectoral level with respect to progress in research and development, have fallen through. There are even fewer grounds for believing that they will be able to play this role at some point in the future. Ministries could only provide for the advance of science and technology by issuing various orders, which is hardly the most effective method, as is found by practical experience. However, there is one efficient way of accelerating scientific and technical progress, namely, to bank on the initiative of major production and research-cum-production amalgamations, by supplying appropriate materials, means and funds, providing these amalgamations with the right to carry on trading abroad, as well as the right to licensing, and also allowing them to be engaged in various forms of cooperation in the fields of science, technology and personnel training. This kind of cooperation (rather than administrative methods) will have to become a major means of concentrating the scientific and technological potential of different branches and of pursuing a coordinated scientific and technical policy. The sectoral management system's control over the bigger part of social resources has resulted in some social functions of ministries and departments becoming inflated. Thus, they are now engaged in many things which are far removed from production activities, such as construction and upkeep of housing, kindergartens and creches, communal and cultural facilities, municipal transport, etc. with their contribution often constituting the larger part. Departmental division of social infrastructure gives rise to less efficient use of resources in the non-production industries. A discrepancy in the availability of resources between "rich" ministries and departments, on the one hand, and local authorities responsible for the integrated social development of a given territory, on the other, prompts emergence of some irresponsible approaches to this extremely important issue. Important differences among departments with regard to the funds allocated for social needs have aggravated the situation by generating different conditions and standards of life inside one territory or even one locality. All this hampers the consolidation processes now under way and gives rise to various obstacles on the way to implementing an effective system of social safeguards.

It is the real transfer of funds for social needs to local Soviets (rather than "persuasion" of ministries and departments) that can help lay the groundwork for reaching the key targets in social policy, i.e. achieving social and economic equality and meeting the population's basic needs. And it is only proceeding from this principle that the differentiated system of wages and provision of social benefits at various enterprises and in different sectors of the economy will be clearly enunciated in social terms and produce some economic effect.

Departments' monopoly rights in this area have the dangerous effect of generating the idea that social and environmental factors are dominated by production needs, while consumers remain dependent on producers. This idea cannot be eliminated merely by developing some inter-sectoral management systems (like, say, directors' councils set up in cities or at territorial-production complexes). Moreover, some sectors and departments may even join forces upholding production and economic priorities at the expense of social ones.

The departmental approach (which is more often than not cited as the chief defect of a sectoral management system) is far from being the only obstacle on the way to the comprehensive solutions of socio-economic problems. It is rather that the criteria applied, which are common to all links in the sectoral system, are limited and tame and this could be identified as a major defect of the departmental approach. Accordingly, the fundamental issue for the further development of socialist ownership is the elaboration of a real, polyhierarchical approach to the use of social resources, a system of equal rights with regard to ownership of socialist property in different forms (amalgamations, enterprises, regions, cooperatives, including consumers' cooperatives, and individual citizens). These owners should be vested with powers and responsibility, which will enable them to balance out the conditions of social and economic development.

An important aspect here is the lack of balance between big, medium-sized, and small enterprises. Evidently, a bigger part of resources (capital funds and labor) is concentrated at some large enterprises. Each one of them subsequently acquires a lot of subsidiaries and service industries. And here the objective law of contemporary production, that is, the concentration of production at big enterprises supported by a network of small and medium-sized ones, comes in operation, though it is implemented in some economically non-effective forms. One of the negative consequences of this is the emergence of "monosectoral" towns, dependent on one or two large enterprises, which largely restricts the opportunities for most inhabitants for choosing their job and life style.

We all need now a thoroughly considered approach to locating medium-sized and small enterprises in the countryside in order to draw into economic use an entire mass of local resources, to develop flexible forms of employment, to draw production and the services closer to the needs of consumers, to develop the economic and socio-cultural potential of medium-sized and small towns.

At the present stage of the Soviet Union's socio-economic development, we also badly need an economic socialist mechanism that would help promote an entire spectrum of economic relations and forms intrinsic to socialism. The structurally harmonized economy, an essential precondition for intensified social production, should also be institutionally balanced out, which would help to establish ownership in the form of an evolving unity of interests of the state, work collectives and individuals.

Collective interests should become a central element in the developing interaction of social, collective and individual interests. They should become a link between the interests of society as a whole and those of individual workers. Collective interests are an historically essential form for augmenting the individual's limited ability to secure his own interests. Through them, the practical essence of economic relations becomes compatible with the thought processes of a particular individual. This particular function of collective economic interests, which plays a leading role in the socialist system of economic motivations, has so far been poorly developed in this country.

The evolution of collective economic interests has proceeded in the form of self-financing activities of production units. Some other forms of collective economic activity, which are linked with consumption, the functioning of territorial communities, the services, and which embrace the entire range of man's economic relations, inducing him to active participation in a number of organized collective efforts, remain poorly developed.

This structural imbalance of collective economic interests is dangerous in that it may be conducive to a weakening of the consumers' economic situation and a contraction of the range of social and economic activities of the people. With no counter-balance on the part of other forms of collective economic interest, the self-financing and departmental interests of enterprises, amalgamations and economic sectors may have become a self-oriented force, restrictive with respect to centralized economic management. Accordingly, while recognizing the need to promote economic independence of local businesses, attention should be focussed on developing all forms of collective economic activity and all collective interests which arise in this connection.

It is important to take account of two intertwined aspects of social-individual ownership (most adequately represented by joint-stock companies), namely, the use of social resources and the accompanying responsibilities. It is the first aspect that obviously predominates in Soviet economic practice. Control by enterprises and industries of resources does not adequately balance out with their corresponding responsibilities, which would safeguard in this context the economic interests of the state, other work collectives, and individual workers. However, it is this system of guarantees which constitutes the social and economic essence of the general responsibility, and which is a fundamental aspect of ownership relations.

Soviet economic experience indicates that both in practical efforts aimed to improve the economic mechanism and in economic research, emphasis was laid on defining the extent of the production units' independence (as well as measures regulating their operation), by means of planned targets, that is, by choosing a definite plan for the disposal of resources, with no due account of the means and methods providing for economic responsibility. The latter was not even considered as a problem in itself. In any case its solution does not necessarily depend on setting forth an "improved" system of planned targets for evaluating the enterprises' operation or determining the "rigidity" of the corresponding planned targets.

The result was that economic responsibilities grew weaker, a demand for production resources swelled (leading to the accumulation of superfluous stocks of resources), incomplete construction projects mushroomed, and economic plans were reshuffled on numerous occasions.

Economic practice clearly shows that the outwardly rigid system of regulatory standards and administrative norms can in no way guarantee the full discharge of economic obligations and is often accompanied by lack of control with respect to local users of resources.

Thus, if measures designed to regulate the economic independence of production units are not supplemented by a developed system of economic responsibilities, they will fail to influence substantially the effective use of resources and facilitate reliable functioning of all businesses.

Extended economic rights of less important self-financing units should be supplemented by a developed contractual form of economic relations, allowing mutual obligations of the participants in socialist ownership to be fixed and providing for the meeting of their corresponding interests if these obligations are fulfilled. The contractual system of economic interaction is a major prerequisite for promoting cooperative administration structures; it should also incorporate various means for protecting the consumers' economic interests.

Another possibility is to extend the rights of workers, adequately to develop collective self-management, to ensure the working people's real participation in decision-making regarding the disposal of property and their direct responsibility for production results. It is also important to preserve and consolidate the social element of property, i.e. labor resources. What we have in mind here is that it is an effective involvement of all the able-bodied people in social labor that constitutes a foundation underlying the general system of social safeguards intrinsic to socialism. The consumption on the part of society of the produce of labor thus constitutes the essence of the processes which regulate the growth and maintenance of the labor element of socialist ownership.

This element's development is inhibited by a number of major problems. The consolidation of the labor aspect of socialist ownership is hampered not only by the fact that some people get unearned incomes in their "classical" form (through speculation, misappropriation, power abuse), but also by that some workers strive to get more from society, while giving less to it, which is, in social terms, very similar to getting an unearned income; and if one considers the scale and the socio-economic consequences of this phenomenon, it will turn out to be even more threatening and dangerous.

While examining the reasons underlying these negative factors, attention is usually focused on the flaw inherent in the existing system of remuneration. However important this circumstance may be, the possibilities for raising the performance of workers are to be found, in our opinion, in socio-economic mechanisms designed to shape an active approach to socialist property; intrinsic to this approach is that there is a social and moral value to assumed responsibilities.

The principle of distribution according to work done is but a component part of the social system safeguarding the "measure" of labor and the "measure" of consumption. This principle is effective only when it rests on some more profound motives and factors related to involvement in socially useful labor and direct control of property.

Central to the system of factors aimed to stimulate better work should be conferring on workers some of the attitudes of a master, and would arise from the workers' efficient participation in management. A delay in development here cannot be compensated by better individual incentives. We should largely abandon (naturally, in a thoroughly considered and balanced-out way) some outdated forms of individual involvement in social labor such as individual piecework, while developing collective forms of labor organization and remuneration, which would adequately correspond to the workers' real involvement in management. It is the work team itself, proceeding from the aforesaid principles, that can work out and effect a differentiated system of wages.

While advancing socialist ownership, typical of which is a real plurality of ownership relations, socialism is in no way centered economically on state (national) property. Unless this becomes clear to all of us and we disperse the smoke screen of ideological attitudes, entailing the mixing up of conceptions of socialism and communism, the motivational mechanisms underlying the socialist economy will operate less effectively than under capitalism. This necessarily will bring about a historical defeat of socialism and reveal its inability to provide a high standard of living. This has been fully indicated by some failures of "socialist" development. Low living standards is too dear a price to be paid for ideological dogmas, depriving society of natural and sound development. Equally groundless is the "nomenclature" conception of ownership based on a "nomenclature" economy: federal subordination, republican subordination, etc. Hence an absurd juxtaposition of federal and republican property, serving, in its turn, as a basis for a totally absurd conception of regional self-financing. In actual fact, however, self-financing is a system of relations applying to an economic entity producing goods of material value, an entity that is not subordinate either to central or regional authorities, but is dependent entirely on the laws of economy, and the market. If we take seriously and without prejudice Lenin's words to the effect that the New Economic Policy (NEP) should be taken in earnest and that it will last for long, we will have to admit that private ownership (in a variety of forms) will remain a major and lasting factor under socialism as well. Without this understanding we in this country will be unable to proceed one step on the path to socio-economic and social progress and will rather aggravate the crisis state of society.

The need to promote the socialization of production, to improve the interaction of social, collective and individual interests in a socialist economic mechanism, to raise the responsibility of all economic units for the results achieved, and to consolidate the labor basis of the socialist property, puts high on the agenda the following issues of the long-term ownership policy:

1. A transfer to harmonized socio-economic strategy regarding the development and location of production, a strategy based on an optimum quota for big, medium-sized, and small enterprises. This strategy also presupposes:

– accelerated construction of enterprises designed to produce inter-sectoral products (first and foremost, in engineering);

– establishment of small enterprises to produce consumer goods, to build cottages, and to launch short production runs of particular spare parts; all these enterprises, also including some service businesses, small retail trade and public catering, could draw on people's money, as well as government financial and credit resources, using flexible forms of employment (e.g. work at home).

2. Intensified socialization in the non-production sectors aimed to turn them into a single complex of territorially organized social infrastructure which would constitute the material and technical basis for the population's sustenance and for a rise in the quality of labor resources. To effect this task it is necessary to

– transfer the task of building the non-production projects to territorial bodies, providing them with the relevant quotas and funds which are now distributed through industrial ministries and departments;

– develop organizational forms and economic mechanisms for the interaction of non-production economic sectors within defined regions, with a view to providing a proportionate and effective redistribution of resources, in accordance with the social development plans for the territories and localities;

– integrate on a stage-by-stage basis the social infrastructures and the transportation system of town and countryside and establish group-based systems of populated localities.

3. Redistribution of control functions with respect to social resources to benefit inter-sectoral, territorial and programmed administration bodies by

– setting up, on a territorial basis, a network of foundations for socio-economic development (local, regional, and federal); these foundations should be financed by means of allocations on the part of enterprises and amalgamations (differentiated according to the regional principle) for their disposal of labor and natural resources; these special foundations should provide for the efficient redistribution of finances among the local, republican, and the federal budgets and ensure the general coordination of territorial development with the measures envisaged in the integrated social programmes;

– raise the role of local budgets, in particular by transferring to them a part of the tax revenue (both direct and indirect local taxes).

4. Vigorous development in all sectors of the economy of various cooperative forms of production and ownership.

The development of collective forms for the organization of production and consumption and the corresponding varieties of cooperative property is indispensable for invigorating and harmonizing the collective economic interests which are to replace the dominion of lop-sided self-financing interests of enterprises and amalgamations.

A broader use of cooperatives in the production of consumer goods, in construction, and the services will help eliminate shortages in some goods and services, and involve in social production some latent reserves of labor and people's savings.

Proceeding from these fundamental considerations, it would be expedient to implement in the near future the following:

– to establish cooperative links between production and research-cum-production amalgamations in various economic sectors in order to launch the production of goods for inter-sectoral use, jointly to build infrastructure projects and to create, for the period of operation of the relevant contracts, a common fund of resources;

– to develop cooperative forms of labor organization and production within enterprises and amalgamations;

– to establish in cities a network of consumers' societies, constituted on a voluntary basis with collective self-administration, to study the population's needs, as well as to trade and to provide services; house-building cooperatives could become one of the forms for development of consumers' cooperation movement in cities; efforts of grassroots consumer societies, including those in the countryside, should be coordinated by district, city, regional, republican, and All-Union consumer societies' councils, vested with relevant powers and disposing of material and financial resources;

– to develop and perfect the cooperative forms of use of the land allocated for collective use, i.e. infrastructural units and recreation areas (family recreation and tourism societies, city and district cooperative holiday-homes, gardening associations, etc.);

– to introduce on a broad scale cooperative and private forms of agro-industrial integration as a basic trend in the development of the agro-industrial complex and a major means of employing cooperative principles in the area of production.

5. Comprehensive development of individual farming, as well as other forms of individual work.

6. Promotion of a contractual basis to socialist economic planning and management, thus turning the contract into the basic means of organizing relations among enterprises, amalgamations and state administration bodies. Raising the status of the economic contract in the system of socialist economic institutions would serve to consolidate the economic responsibilities of all economic units and turn it into a central element guaranteeing and legally substantiating the parties' obligations, and underpinning economic and social development plans.

While clearly delimiting mutual obligations and responsibilities and securing within these limits the complete interests of the state, enterprises, and individual workers, the system of economic contracts also allows clear definition of and subsequent expansion of the area of activities, in which the economic interests of work teams are maintained by their own economic initiative, with the resulting risk being incumbent upon the relevant enterprises and amalgamations.

7. Improvement of socialist distribution mechanisms, thus creating the corresponding preconditions for intensifying and raising the efficiency of social production.

What we have in mind here is primarily a consistent demarcation of social and economic functions with respect to an aggregate profit. Wages and other forms of labor remuneration should no longer be defined as a set of social safeguards guaranteeing minimum living standards. Any kind of ministerial consumption funds should be eliminated. Inter- and intra-territorial differences in the size and quality of social benefits intended for collective use (and allocated from social consumption funds) should be substantially reduced.

Given a developed system of social guarantees, it is essential, in order to stimulate better work, to

– link remuneration and performance at work of individuals and of entire work teams;

– establish an effective system of personnel management, which will also include planned release of labor resources and their training; heads of enterprises should be released from the need to re-employ released workers, with this function being transferred to some special federal and local agencies;

– extend the rights of top managers of enterprises and of work collectives in establishing rates of pay.

The promotion of socialist ownership thus lays the groundwork for perfecting the whole system of socialist production relations, including the relations and forms of socialist distribution. It also embraces key aspects of shaping the socialist mechanism of economic development and stimulating better performance at work.

All-round development and consolidation of socialist ownership presupposes an effective system of taxation at all stages of social production. A plurality in ownership relations is increasingly becoming the foundation of political plurality and vice versa.

6. Socialist-Based Motivational Mechanism for the Effective Use of Resources

Vladimir I. Lenin said on numerous occasions that higher labor productivity (to be more precise, an aggregate efficiency in the disposal of material, labor, and natural resources) is indispensable for the victory of a new social order. At the present time, however, the efficiency in the use of all production resources is much lower in this country as compared to industrialized nations in the West. The Soviet Union is also at a disadvantage if we consider the motives and incentives underlying the effective disposal of resources. If the situation remains unchanged, socialism will fail in economic competition with capitalism, will fail to secure a higher level of economically efficient production. There is no room here any longer for all sorts of lullabies to the effect that, socialism is less effective economically than capitalism, but it provides the people with an entire range of social benefits, that is, socialism is more effective in social terms. To switch off such lullabies, one does not need to produce any profound theoretical generalizations; it would suffice to look at some statistical data to become aware of the fact that socialism has been slowly and continuously losing its competition with capitalism.

The time has come for serious and adequate evaluation of the economic realities. The search for truth should become the aim of all research, including that done by social scientists.

The motivational mechanism underlyirtg the use of resources is a profound conception, which cannot be easily reduced to any index (e.g. the remuneration system), however important it might be. The remuneration system though, plays a leading role here, because it is organically linked to other elements of the motivational mechanism. Let us give a brief account of them all.

1. Remuneration system. From one according to abilities, to one according to work done. This constitutes the essence of the socialist idea, but neither of the two goals mentioned in this principle has ever materialized in the development of socialism. In actual fact, it has not yet been able to unleash man's full potential or to remunerate him duly for the work alone. The visible development of Soviet economy has been running counter to this postulate. Extensive economic development, combined with continuous shortages prompted an artificially high demand for labor (to the detriment of efficiency). Some monopoly elements come to the forefront at the labor market, which deformed the distribution principle by establishing egalitarian distribution and destroying the connection between the measure of labor and the measure of consumption. Wages were vested with some social functions, and they were adjusted to fit to socially admissible standards. Individual's potentials were increasingly regarded and the principle of remuneration "according to one's work" was violated on a large scale. All this caused economic, social and political corrosion. Thus, remuneration according to work done did not act as an effective incentive for the efficient use of material, labor and natural resources, which brought in its wake the lowering of socialist economic rates and the standard of living. This was a continuous process, which proceeded despite some constant but futile calls to cut it short. A profound restructuring of the entire interdependent economic system was needed, but this, it was claimed, was fraught with troubles and difficulties, and so all remained as it was.

One of the topmost requirements for the radical restructuring of the economy is a full revival of a system of incentives to stimulate performance at all economic levels. However, we should not turn a blind eye to existing realities. The renascence of the principle "from each according to his abilities, to each according to his work" would demand a tremendous effort on the part of the whole society (we will elaborate on this below). Economic incentives could only become effective when combined with full, economically sound, and socially reasonable employment, rather than with artificial full employment. A worker should be economically induced to struggle for an adequate standard of living. The scientific and technical revolution, manifested in economic production efficiency, may bring about a mass release of labor. If this matter is approached in earnest, a set of constructive measures should be elaborated to regulate employment and to set up a system of training. Complete and economically sound employment might be accompanied by technological and structural unemployment, which can be eliminated by means of retraining and stable rates of economic growth. This temporary "unemployment" will obviously induce the workers to make a more effective use of resources, thus drawing a line between social philanthropy and economic expediency.

2. Working conditions and the character of labor. Mention should be made in this connection of an objective economic law which says that the higher the rate of real profits, the higher the stimulating role of such factors as the character of labor and working conditions, involvement in decision-making, etc. The influence of working conditions and of the character of labor will have to play an increasingly important part in promoting the motivational mechanism of socialist economy, though, of course, this influence will remain different for different population groups. In the Soviet Union, the situation in this area is far from favorable. Over a million people are engaged in manual labor. A few million people are engaged in arduous, unskilled and sometimes harmful labor. It is particularly distorting in social terms that millions of women are employed at these jobs. A few jobs possess some creative and attractive features, but the unpleasantness of most is due in large measure to low gross output, antisocial aspects of production operation, inadequate capital investment to eliminate unpleasant jobs. Even the gross output indices adversely influence the growth of funds aimed at improving the working conditions. This negative influence cannot even be compensated by growth in the people's real incomes, which has considerably slowed down. Regarding this factor (and the two considered above), the Soviet Union is at a disadvantage compared to the West. So far there have been no signs that the tendency is changing for the better.

3. The system of social safeguards. It may seem strange, at the first sight, that this factor is included in the motivational mechanism underlying the effective use of production resources. But only at first sight. Egalitarian attitudes reigned in this country for decades, and we often explained this by the fact that a differentiated system of incomes, corresponding with the principle of distribution "according to one's work", may have resulted in some unwarranted stratification of society, social tensions and other troubles, which have often been used to justify various unsound measures. A developed system of social safeguards may also serve as a basis for economic decisions. In practice, a socialist society need not fear some economically justifiable stratification as regards the size of incomes, basic needs being socially protected. Due account should also be taken of the fact that, through the system of social consumption funds, various benefits are granted to the people and their educational standards are upgraded, which lays the groundwork for raising the efficient use of production resources. Regretfully enough, the development of this aspect of motivational mechanism under socialism (contrary to a widespread view on this matter) is still lagging behind as compared to some industrialized countries in the West.

4. Workers' relation to property and their involvement in decision-making at all levels of economic hierarchy. In this area socialism may outplay capitalism, given that socialism's economic, social and political image has been radically changed. A markedly greater participation of the people in the management of property and social production remains no more than a dream, a kind of an "ideal" to be pursued in the future, but, alas, far removed from real life. It is really impossible to cut the Gordian knot of these problems at one stroke, especially if one considers that for decades everything possible was done in this country to alienate the workers from managing property and production; nothing was done to develop in them the sense of ownership in economic, social and political terms. On the contrary, they were "purposefully" led in the opposite direction under the cover of political declarations. Yet it is here that socialism may have an advantage over the West. Accordingly, it is expedient to secure a fundamental change in this area, otherwise the Soviet Union will have to admit that the motivational mechanism of capitalism is stronger than that of socialism with all the ensuing consequences. We should all get rid of our illusions, and stop recognizing some potential "advantages" of socialism none of which has so far materialized. Naturally, leasing, cooperative work teams, family contracting and the like can only be welcomed, but we cannot lease the whole of social production, including the work force. As noted above, the relations of ownership constitute a fundamental economic, social and political problem, without solving which we will be unable to prove socialism's solvency. Political system, cultural standards, and religious aspects should also be considered in this context. Not all realities can be explained by analysis of concepts of economic basis, production and economic relations (as Engels warned us). Without taking due account of his warning, we will just turn into vulgar economists.

Chapter IV
The economy at a crossroads

Dmitry S. Lvov

Corresponding Member of the USSR Academy of Sciences

Sergei Yu. Glaziev

Head of Department, Central Economic and Mathematical Institute, USSR Academy of Sciences

1. The Economic Reform and Its Tendencies

To transform the command economy into a market economy is a task of exceptional complexity, which could hardly be solved without social tensions within a short time. The five years of economic reform in the USSR have brought none of the expected results in introducing market institutions and enhancing efficiency of public production. Instead of the anticipated acceleration of economic growth we are currently witnessing a disorganized performance of state enterprises and a disastrous decrease in the output of many products; certain sections of the population have faced a fall in their real incomes, with the prices for consumer goods noticeably and steadily rising. The consumer market is in the throes of acute shortages. Many commodities which until recently were in abundance have disappeared from the shop counters.

These mounting economic difficulties are to a great extent due to the chosen course of economic reform, which was predetermined by the ideological concepts of appropriate ways to reform the socialist mode of production and erroneous doctrines on marketing patterns.

The readjustment of the national economic management system, launched in 1985 when a profound crisis in public production came to be fully apprehended and identified, had by 1990 passed through two subsequent stages. These were, in fact, searches within the confines of conventional guidelines for improving the socialist economic mechanism.

The first stage of economic reform, which began with the adoption of the Law on State Enterprise and the endorsement of a package of decrees passed by the CC CPSU and the USSR Council of Ministers on July 17, 1987, continued up to the approval by the Second Congress of People's Deputies of the USSR of the government's Programme for implementing economic reform in

November 1989. This stage was characterized by reform along the conventional guidelines of the command economy, and by more independence for enterprises in the matters of pricing, logistics and profit distribution while keeping up their administrative subordination to sectoral ministries and the command assignments of superior administrative bodies, though these assignments took the shape of government orders. Further, the power of governmental and Party bodies in economic matters was weakened, and new opportunities opened for individual and cooperative activities. The system of state planning and allocation of resources remained virtually intact, despite the declaratory provisions of the respective decrees. Though the Law on State Enterprise provided for the election of production managers with subsequent approval by a superior body, the existing procedures in planning and logistics kept them subordinate to the ministries.

The disbandment of some sectoral ministries did not bring abolition of the respective production-departmental systems which were reproduced in the form of associations and companies incorporated in the old institutions of national economic management. The readjustment of a number of sectoral ministries into associations of enterprises and incorporation of several groups of enterprises into companies, while preserving the command system of economic regulation, in fact, only complicated their management by the central economic agencies. The situation was further aggravated by the merging of several ministries and by staff cutbacks in the ministries and departments, including the central ones. With the agencies responsible for control of publicly owned production remaining virtually intact, the changes made during the first stage of economic reform resulted mainly in a weakening of the centralized national economic management, with the command economy preserving its basic control economic agencies, and in strengthening the economic power of state enterprises under the unaltered mechanism of their economic responsibility. Along with the weakening of the central economic departments, this led to slackening of production discipline throughout the national economic management, acute shortages and further imbalances between commodities and money, a state budget deficit and growing inflation. Also there were fewer opportunities for conducting a vigorous technical and economic policy and implementing the advanced structural shifts in public production.

In other words, the first stage of the economic reform, contrary to the intentions of its initiators and in spite of the abolition of certain economic departments, brought nothing but consolidation of departmental principles in management of public production and, as a result, a further decrease in its efficiency and its aggravation of structural anomalies, manifested as mounting shortages of production resources, frequent and critical shortages, and foreign exchange deficits with a simultaneous worsening of the national currency balance. The weakening of the central economic departments meant that inadequate decisions would likely be taken by the supreme governmental bodies which became an object of direct lobbying by the production-departmental systems. The latter found a route for breaking into the supreme governmental bodies through their standing agencies (the bureaus of national economic complexes) at the USSR Council of Ministers, bypassing the central economic departments which had previously settled interdepartmental disputes and guarded the supreme governmental bodies from taking unilateral decisions liable to disrupt to national economic balances. This weakened central control over departmental claims could have no effect other than to aggravate the national economic disproportions and the general economic situation. This, in its turn, stimulated the further unfolding of economic reform.

The second stage began with the approval by the Supreme Soviet of the government's programme for intensifying economic reform. The main guidelines remained the same, but it was decided to launch some radical transformations.

The decisive factor in expanding independence and self-determination was, as before, to reduce the administrative dominance of ministries and local Party bodies. During the first stage an attempt was made to transfer control of production management from the superior governmental bodies to the work force, and to give the latter control over strategic decisions (production plans, profit distribution, etc.), and the administration and power to elect their managers. Thus, the Law on Socialist Enterprise vested the work teams with the right to approve their chief executives. The newly adopted fundamentals of Soviet legislation on leasing allowed a work team to lease a state enterprise if two thirds of the workers voted for it. In this way the work teams of state enterprises have been granted the key proprietary rights to dispose of the public means of production.

Let us consider the probable effects on national technical and economic development of wider independence of state enterprises as a result granting proprietary rights to the work force. First of all, it gives rise to a new governing center at an enterprise, such as the work team council or, a more recent innovation, a council of leaseholders, both of these bodies having more power than the management though not having their executive skills. Being accountable to the general assembly of its employees and to the work team council, the management has to confine itself to decisions which would be in the interests of the majority of the personnel. Under the command economy most working people were deprived of the opportunity to take part in decision-making, and this estranged them from their work and their enterprises. The cumbersome system of formal administrative control (inherent in most state enterprises with their linear-functional structure of management) failed to stimulate their personnel into regarding their work as personally rewarding and socially profitable (or profitable for the organization). Therefore, most people working at state enterprises seek higher wages for less work.

The ways people are motivated cannot be changed by introducing new rules into human activity. In this case certain proprietary rights accorded to the work forces of state enterprises do not change the actual conditions for most of their members, but the latter do get an extra chance to put additional pressures upon the management. The last two years of practicing individual democracy of this kind have shown that such chances are used by the workforces primarily to get higher wages. This tendency manifests itself the amount paid in wages growing more rapidly than labor productivity; the former increased 3.4 times in the last year alone.

The right granted to the work team to take strategic decisions makes the management redistribute profits in favor of consumption to the detriment of the long-term interests of production development. The latter consideration is especially important, since it contributes to unfolding economic contradictions and affects technical progress in the national economy.

The conflict between a production management responsible for efficient performance and a workforce interested in higher wages is a well known aspect of labor relations. It exists irrespective of the owners to whom the management is responsible, be it shareholders, a bank, a state body or a private owner. The location of the governing center outside the business and sphere of influence of its workforce induces the management to strive to increase production efficiency despite the desires and, as a rule, regardless of personnel resistance. In the economic conditions outlined in the Law on State Enterprise, the draft Law on Socialist Enterprise and the adopted fundamentals of legislation on leasing, the management of a socialist enterprise is responsible solely to the workforce. The Yugoslavian example, as well as the latest scientific knowledge on patterns of social behavior bear out the inadequacy of such control mechanism.

First, in the minds of workers the motive of higher wages for less intensive labor usually predominates over the motive to contribute to production development.

Second, concentration of control over production efficiency within an enterprise very clearly consolidates its authority in the economic sphere. The production interests merged with the interests of the workforce acquire political connotations and manifest themselves in the pressure exerted by the working people on local and central governmental bodies to grant exclusive advantages and privileges to their enterprise.

Third, if right to strategic decision-making is delegated to the workforce, innovations which are invariably accompanied by changes in work patterns are slowed down, by the resistance of the workforce. The introduction of innovations involves overcoming this resistance which entails quite a number of contradictions between the management and the workforce of an enterprise.

Consolidation of its authority in the economic sphere would hardly enhance the innovatory potential of state enterprises and the technical progress of the national economy.

So, public production based on autonomous enterprises has drawbacks reflected in lower innovatory activity and productivity which impede the raising of public production efficiency.

Theoretical considerations and practical experience bear out the thesis that a highly efficient market economy could not be based on collective forms of ownership. The efficiency of the market economy is known to be determined by free redistribution of resources from lesser to more effective economic spheres under the pressure of market competition in commodities, labor and capital. In all of these markets freedom of economic activity and the overall redistribution of resources is more limited in collective than in private and even in state enterprises.

Changing a production programme under the conditions of competition in the commodity markets calls for technological and organizational innovations, very hard to introduce in the conditions of collective ownership, for the management is limited in its attempts to enhance production efficiency by the requirement to satisfy the desire of the workforce to keep up wages and to maintain relaxed working conditions. Collective enterprises respond sluggishly to changes in competition and, generally, are driven out from the dynamic sectors of market economy by corporate and private capital which is more mobile and less subject to pressure from the personnel.

Collective enterprises are also characterized by having less mobile man-power resources. Labor mobility among enterprises and sectors in the conditions of collective ownership is quite limited, both because the employee is bound to the enterprise at which he is working because it is his property and also because the management is not so free to discharge redundant labor. For these reasons the labor market under the predominance of collective forms of ownership in the economy is imperfect.

The same applies to the market of capital, for its redistribution among en-terprises, sectors and economic spheres implies free withdrawal and investment of money, and the existence of institutions involved in capital circulation. In this context of collective ownership, circulation of capital is hampered by its low concentration, its dispersion among numerous owners, and their lack of interest in higher profits. Redistribution of capital to more effective businesses is checked by the determination of workforces to continue production regardless of efficiency. Finally, collective ownership excludes any possibility of setting up a stock exchange which, being the market for the owners of the means of production, is the basis of capital markets in the advanced countries. Without it, an effective system of credit is impossible, as it provides the only reliable system of indicators of the market value of various enterprises and allows the securities market to function. As we can see, public production based on collective ownership has a number of fundamental defects which make it impossible to shape a highly efficient market economy along these lines.

Due to the motivations of the workforce above mentioned, an economy based on collective ownership is oriented towards consuming the profits from economic activity. It is notable for underdeveloped institutions responsible for productive accumulation and for a constructive resolution of the contradictions implicit in the market economy, also for integration of self-sufficient businesses aiming at a higher efficiency with rational performance of public production. The contradictions that arise from a market economy in which collective ownership predominates, are expressed as growing inflation, wider disparities in the wages of people employed at different enterprises in different regions of the country, and in disorganized public production.

It is only natural to wonder why such a strange and obviously inadequate choice of process for economic reform was made by the Soviet government. The answer rests in the social context of reform in the USSR. It is characterized by a quite noticeable social resistance to the introduction of a market economy and to integration of this country in the world market and this is a consequence of the technological backwardness of many sectors of the national economy.

The system of command management in public production provided no stimuli for the employees to work hard and made them accustomed simply to collect their wages regularly rather than to strive for higher incomes. Managers of enterprises, being subordinate to administrative departments, also lacked interest in raising either production efficiency or labor productivity and were mainly preoccupied with the prompt implementation of command assignments which made them conceal the production potentials and accumulate reserves, including labor resources notwithstanding the production costs. In their turn, sectoral ministries which under the command economy hold the most power behave as large organizations might be expected to do, and manage their industries simply according to routine. In the absence of competition, ministries, likewise their subordinate enterprises, strive for maximum reserves through bargaining with the central economic agencies for excessive quotas of resources, to fulfil their plans.

It was only natural that operation and proliferation of the national command economy eroded labor motivation in people employed at state enterprises, widening the gap between the intensity of labor, its quality and social profit, on the one hand, and remuneration for labor, on the other. State enterprises came to have labor redundancy which, according to the estimates of managers of many enterprises, now amounts to 20–30% of their staffs.

The customary parasitic attitude is not the only obstacle to making wide sections of the population adapt themselves to market relationships. Another, no less major problem is the widespread stereotypes in the attitude of people to marketing.

For many of our compatriots a market economy and competition are invariably associated with frauds, financial speculations, tricking the customers and sweating of labor. The popular saying "No tricks, no bargain" aptly reflects the general attitude to marketing.

In actual fact, the present-day realities of advanced countries with a market economy differ noticeably from the popular notions of business anarchy, impoverishment of working people, cheating of customers and rivals, and other similar attributes of the capitalist world which have been inculcated into our public consciousness for decades. The age of robber barons, greedy, merciless and dishonest capitalists, workhouses and the like, which the mass consciousness associates with a market economy, belongs to the past. Nowadays a prosperous businessman, as a rule, has a stainless reputation as an honest and noble person; companies fear like death any reproaches for adulterating the quality of their products and for cheating customers or governmental bodies; humane relationships and elements of industrial democracy have become indispensable to company administration. Regrettably, an inadequate approach to the socio-psychological aspects of the modern market economy engenders not only philistine prejudices against economic reform but also shapes the actual market relationships in a corresponding way and, moreover, underlines numerous strategic miscalculations in the strategy of its implementation. The private sector in the Soviet economy today is in fact reminiscent of the market economy as it used to be caricatured in mass propaganda. Bad quality, gross cheating of customers, malpractice towards rivals, and similar features of the current cooperative and market trade naturally exasperate broad sections of our population and associate in the public consciousness business initiative with swindling.

While appraising the difficulties involved in the formation of market relationships in the USSR, the following considerations should be taken into account, that side by side with the problem of unbalanced economics there exists also a considerable psychological unpreparedness of the majority of people to work and survive in the conditions of the market economy. We should perhaps go through several intermediate stages in creating the ethics of business relations. This might involve forming a new image of the "hero of our time", as a creative, entrepreneurial individual, in the mass consciousness. For new Soviet businessmen it is very important to understand that today it has simply become unprofitable to engage in dishonest dealings and to cheat partners, customers or workers. Any fraud sooner or later gets known to the public while the loss of the reputation of an honest and noble minded businessman shuts the doors of banks, business clubs and associations in the face of the accused while customers turn away from companies with a bad name. Good reputation has acquired an exceptionally high commercial value.

A market, in effect, involves an aggregate of rights. The rights of producers to take independent decisions on the range and volume of production, to fix prices and to enter into various contractual relationships with partners in economic activity. The rights of customers to obtain goods of the desired quality and to choose their own suppliers. The rights of working people to be paid for their labor in accordance with its quantity, quality and their skill, and the rights of the population at large to keep up acceptable living and income standards and to have their interests protected by the government. The rights of the state to improve taxes and to implement whatever decisions are necessary to fulfil its functions.

All these rights exist not on their own but in a complex system of social relationships regulated by legislatures. Without a developed system of institutions to protect the rights of producers, customers, the population and the state there could be no efficient market. Since the efficacy of legal institutions is determined by the general levels of legal education and the ethics of business relationships, a sufficient degree of mass legal education and business ethics is indispensable for efficient marketing.

Indeed, market traders are shamelessly bargaining for maximum profit. To gain it they choose the simplest and most obvious ways. Exorbitant prices, speculation, adulterated products, tax evasion – these means of making profit are, of course, much easier than raising the efficiency of production. The urge for maximum profit pushes businessmen into unfair competition and only an advanced system of public control and the strict demands set by the business community for probity and professional ethics, propped up by legal means, keep them from actions which run counter to the public interests.

At the same time, working people are interested in higher living standards regardless of the quantity and quality of their labor. Only high standards of production relationships backed up by an advanced system of coordinated interests and a legitimate settlement of contradictions can guarantee economic growth against pressures for excessively high wages.

The significance for an efficient market economy of high standards of legal education for the population and of business ethics becomes more apparent if we compare market relations in advanced and underdeveloped countries. In Latin America, for instance, there operates virtually the same system of production relations as in the USA, Western Europe and Japan. The results, however, are quite different. And the matter is not the size of the accumulated production potential (in West Germany and Japan it was nearly destroyed by the war forty years ago). The main reason for low efficiency in the Latin American (even more so in the African) market lies in the low quality of legal education and of the ethics of economic relationships in the public consciousness of these countries. Their economy is not protected by public control institutions against the arbitrary rule of local and foreign businessmen. For this reason business dealings in these countries more often than not have the piratical form which gives rise to social upheavals. Protection of business enterprise, overcoming of social tensions and the maintenance of stability are executed in such conditions, as a rule, by military dictatorships.

Low standards of legal education and of business ethics make it customary, in the context of market competition, to resort to such means as bribery, blackmail, violence and, in some cases, to overthrow of the state. A lack of institutions for state and public control of business leads to unfair competition of producers, large companies in particular. Outrageous prices for finished products and ridiculously low prices for raw materials and labor, adulterated goods and favorable conditions for business gained through bribery, are all well known phenomena in the economies of developing countries.

At the same time, poor labor relations prompt working people to seek better wages not, say, by raising productivity, but by going on strike and by engaging in political struggle to raise their income irrespective of the real productivity of social labor.

Introduction of market relationships to a society with poor legal education and undeveloped business ethics brings about uncontrolled inflation, production anarchy, racial conflicts and economic and political instability in general.

Regrettably, the legal education of broad sections of the Soviet population is still at a low level. People here have long been accustomed to judge things not by law but by conscience (i.e. in conformity with the conventional notions of social justice). The seven decades of Soviet government did little to cultivate legal consciousness and ethical labor relations. Nevertheless, the relatively high educational and general cultural level of the greater part of this country's population make for hope that these attributes could get stronger alongside the restructuring of the economic and political system within a sufficiently short period of time. However, it would take at least twenty years; at any rate, such a radical transformation in the public consciousness is most likely to occur during the lifetime of the next generation of Soviet people.

Perestroika would succeed provided a fundamental change in the public consciousness takes place. Until the bulk of the population learns to think and act in a new way, no transformations, no matter how radical, in the system of national economic and social management could be guaranteed to be irreversible. Therefore, introduction of new economic relationships should be accompanied by the mass cultivation of the corresponding system of values.

Among other things, respect for property is indispensable for business enterprise and market relationships. This legal rule is extremely undeveloped in the public consciousness. Until recently, private ownership of the means of production in combination with hired labor remained unlawful and, therefore, open to illegal practices on the part of both individuals and the government. The mass-scale organized crimes against the property of cooperatives which have not found wide public censure are another side of the same phenomenon. To all appearances, a lack of respect for one's own property and that of other people, which is a remnant from communal consciousness and barrack-like socialism, would greatly hamper the development of market relationships. It could not be healed by severe legislative acts or intensive propaganda. The best medicine in this case is to enlarge private property of broad sections of the population and to encourage all kinds of work and enrichment of our citizens.

The reverse side of an indifferent attitude to property is estrangement from work. As was mentioned above, over the long years of the command system our people lost the sense of relationship between the quantity, quality and skill of labor, on the one hand, and the size of payment for it, on the other. Research studies have revealed that the growth in wages at industrial enterprises depends neither on the efficiency of their performance nor on raising the labor productivity. Psychologically the working people got accustomed to collecting rather than earning their income, even if it is a certain guaranteed minimum. Hard work to get bigger earnings is looked down upon as something indecent and improper. It is no accident that broad masses of the population regard high incomes as something illegal.

These features of the mass consciousness, shaped over more than one decade, reflect the status of the individual in the command system, which was that of a person with no need of initiative and enterprise and merely striving for the means of subsistence. This status finds its most graphic manifestation in the citizen's attitude to the state.

The public consciousness views the state as a patriarchy, a strict and demanding master standing over society, running all public affairs and supervising a fair distribution of profits. The people feel themselves to be employed by the state and paid wages for their services.

This attitude to the state is in striking contrast with that in an advanced market economy, considering the state as a common burden, a heavy load shared by everybody as a civic duty. Since the state with its bulky machinery is functioning thanks to its citizens and to businesses which finance government expenditure from their own incomes, the upkeep of the state is considered to be a civic duty which is necessary for the unity and security of society to be upheld. A popular view is to compare the state with a night-watchman who had to be hired, to keep order.

The position of the state in society is manifested in the way its expenditure is financed. In the command economy it is customary to assign part of the profits made by state enterprises to the state budget under departmental statutory enactments. It is noteworthy such revenues return the same way, albeit in reallocated form. In this way, profitable enterprises pay for the wasteful ones, as in a patriarchal family where healthy members keep their dependents.

The citizens also finance the state, with 13% of their wages regularly withdrawn into the state budget. And again it is withheld but not given by the citizens on their own, with the dependent status of a citizen as servant of the state being engraved once more in the mass consciousness. At the same time the citizen gets from the state free education, free medical care, virtually free housing, and essential goods at over-subsidized prices. All this is done by civil servants and employees of the distributive networks as in a communal state with its gathering and distribution of the public product.

Destruction of the patriarchal-governmental institutions responsible for production and distribution, which will inevitably come with the market economy and with democratic political institutions, deprives broad sections of the population of the customary forms of social protection. A market environment makes man display completely different adaptations than command-based production and social life. Thus, a person accustomed to guaranteed employment may prove absolutely unable to adapt himself to unemployment, even under the appropriate social security system. It is neither so easy to pass from life in conditions of stable and low prices, with its scarcities and queues, to a life in the conditions of differentiated and fast-changing prices, attended by the devaluation of money and of fixed incomes. It is equally difficult to make people pay for the services which they were accustomed to receive free or nearly so, especially when these services belong to the essential category. All these transformations may arouse social tensions and make people act inadequately.

The public consciousness is not prepared for basing the national economy on market principles and this accounts for numerous setbacks suffered by economic reform during its first two stages. The policy of setting up the so called socialist market on the principles of collective ownership was not only due to the government of the time being unable to base national economic management on the market relationships, or to Soviet economics having only a poor understanding of patterns of the market economy. It was, in fact, predetermined by the ideological doctrines and stereotypes dominating in the society and clearly manifested in the key trends of our socio-economic thought.

As might have been expected from knowledge of how collective ownership operates, and in accordance with the theory of a self-managing company, expanded self-government of workforces brought about redistribution of income of state enterprises into the wages fund, with increase of inflation and mounting imbalances in public production.

It should be called to mind that the Soviet economy is notable for an extremely high concentration and monopolization of public production. The overwhelming part of industrial production is concentrated at large enterprises employing more than 1000 people. According to the available data, in each sector nearly half of the national output is produced by one or two enterprises. Thus, the mechanical engineering monopolists produce 80% of the total output.[1] The introduction of market relationships based on collective ownership under such conditions incurs great risks of a rapid rise in prices, catastrophic reduction of production and a fall in labor productivity.

Without understanding the conventional technological structure of the national economy and its intrinsic disproportions it is impossible to make prognoses as to the implications should the tendencies which became apparent during the first two stages of economic reform be retained. Neither can suggestions be made on what alternative market institutions might be introduced. This, in its turn, requires knowledge of the patterns inherent in technical and economic development.

2. The Mechanism of Development and the Technological Structure of the Soviet Economy

In recent years world economics came to treat economic dynamics as an uneven and uncertain process in the evolution of public production. From this viewpoint technological progress is perceived as a complex interaction of various technological alternatives carried through by competing and cooperating economic entities in the conditions of the corresponding institutional environment. The choice of alternatives and their implementation in the form of structural changes in public production are realized through the complex processes of training and adapting a society to new technological potentials. These processes are mediated by multifarious non-linear positive and negative feedbacks which determine the dynamics of interacting technological and social transformations.

The new approach to evaluating the control of technological progress calls for a new interpretation of the economic structure. It is of vital importance to choose such an approach to the economic reality which would secure stability of its elements and their interrelationships in the process of technical development. The concept of economic structure which would be adequate to this task involves selection of a basic element that could not only retain its integrity in the process of technological shifts but could also be the carrier of technological changes, i.e., this element should not be broken down further in analysis of the changes involved.

For this basic element we suggest an aggregate of technologically related production, which would retain its integrity in the process of development. By means of a technological network of the same type such aggregates are united into a stable, self-reproducing entity, a conglomerate of related production,

---

[1] V. Gurevich, "The deadlocks of total monopolization". Ekonomicheskaya Gazeta, No. 13, 1990.

making up a technological structure. The latter embraces a closed production cycle, from extraction of natural resources and vocational training of personnel up to unproductive consumption. This approach to the technological structure of the economy permits description of its dynamics as a process of development and succession of technological structures.

Each technological structure is characterized by a set of basic products, a group of leading industries, a sector of unproductive consumption, personnel skills, etc. and rests upon definite types of energy supply, structural materials, means of transport and communications. In other words, the stage of technological progress is derived from the development of production in the corresponding technological structure. The stages of technological progress (and the corresponding technological structures) succeed one another. The previous stage shapes the material and technical basis of the next one. A new technological structure emerges from the depths of the old one and in its continual development it adapts the previous stage of technological progress to the requirements of the technological processes which are at its core. The life cycle of a technological structure comprises four phases – formation, growth, maturity and decline – and has a characteristic form of two pulsations (Fig. 1). The first, "small" pulsation corresponds to the phase of formation, when expansion of the components takes place in an unfavorable economic environment owing to the predominance of the previous technological structure. During this phase a technological structure is limited in its development owing to the relative inefficiency of its technologies and the resistance of economic organizations and institutions of the previous technological structure. As the growth pattern of a new technological structure takes shape and the corresponding institutional changes are introduced, rapid expansion of the new technological structure becomes possible, with the form of the second, "big" pulsation.

The co-existence in the economic structure of integral complexes of related production makes technological progress uneven. Contrary to the widespread simplified notions of technological progress as a continual process of improving public production by means of "washing out" the outdated and introducing new products and technologies, technical and economic development is in actual fact based on alternating stages of evolutionary improvement with periods of economic structural reconstruction in the course of which a complex of radically new technologies is introduced.

In the market economy the formation and replacement of technological structures as part of economic processes are manifested as lengthy waves of the economic situation. The rates of economic development change, depending on the current phase of a technological structure – formation, growth, maturity or decline. Being on the ascent in the phase of formation, these rates reach their maximum in the phase of growth, after which, having exhausted the potentials of the production component and having satisfied the corresponding social

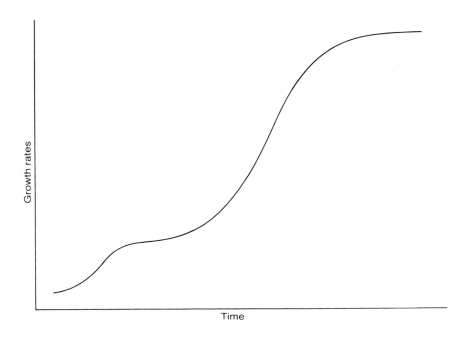

Fig. 1. The life cycle of a technological structure (technological wave).

requirements, they begin to descend to their minimum in the phase of decline. During this last phase the sharply decreasing profitability of capital investments in conventional technologies makes it necessary to introduce radical innovations, which lay the foundations of a new technological structure. The latter starts a new cycle of the wavelike change in the economic situation. Through the mechanism of market self-adjustment innovations and shifts in the key economic sectors, such as mechanical engineering, production of structural materials, raw materials, power and civil engineering, transport and communications, are synchronized. Synchronization is based on technological interdependence. Innovations stimulate and complement one another. Breakthroughs and inventions pertaining to only one of these sectors remain unrealized or, at any rate, find no due currency until the corresponding innovations in other sectors emerge and the requisite conditions for developing the entire system of related production emerge. Nevertheless, the products of one and the same technological structure reach their maturity and the limits of their growth more or less simultaneously, since they have completely satisfied their common market and have exhausted the potentials of their technological networks.

Replacement of technological structures in a market economy is due to the falling profitability of obsolete products caused by falling prices resulting from the satisfaction of the corresponding social needs and exhaustion of their technological potentials. The moving force behind this mechanism is competition among businesses which forces them to look for the most effective investment of their resources and which is channeled into promising directions by the institutions of centralized regulation. The latter should be taken into account while considering rational ways for putting economic reform into effect.

The present-day market represents a highly organized community of private, governmental and public organizations. In contrast to the theoretical model of the competitive economy, it is characterized by stable and long-standing cooperative relations not only among suppliers and consumers, private companies, banks and government institutions, but also among competitors engaged in production of similar goods and services. The market today is distinguished by cooperation and an established hierarchy of various economic entities as well as by their competition.

The current stage in market economies is characterized by establishment of stable horizontal relations among economic subjects rather than by consolidation of vertical hierarchical structures. Ineffective giant corporations with their bureaucratic administration are giving way to flexible and closely interrelated incorporations of big and small, self-sustained and dependent companies, banks, non-profit institutions, higher educational establishments, laboratories, research centers and government organizations. Monopolization with a view to excess profits is replaced by close, mutually advantageous production and scientific and technical cooperation among competitors seeking to survive in international markets through rapid technical development. At the same time state interference is chiefly focused on encouraging highly-effective marketing to solve the problems of social security, education and vocational training, to set up and maintain the capital-intensive transport and information infrastructures and to secure progressive technological shifts in the economy.

The latter problem is currently given particular attention in the advanced market economies. More vigorous efforts of states to encourage technological progress are due to the mounting comprehension of the decisive role of technical development of public production in ensuring the national economic and political strength.

In many cases the state regulation of structural changes in the economy of advanced capitalist countries has been highly effective. For instance, the most noticeable structural shock in the postwar years, the so called energy crisis of 1973, was in fact prepared by the joint actions of the American government and a small number of the largest U.S. corporations. One of the key aims of these joint efforts was the mass-scale introduction into the American economy of a new technological structure based on energy saving through using innovations in microelectronics. These innovations were virtually ready for mass-scale introduction into American industry but were held back by the unfavorable pricing imposed by the previous technological structure. The sudden change of price proportions and the immediate response of American customers and producers that followed, prepared the ground for a miraculously rapid introduction of microelectronic technologies and the attendant basic products of the new technological structure.

Despite the great variation in ways of combining market and centralized regulation of economic activity aimed at accelerated technological progress, certain general features can be identified in all advanced countries. First, centralized regulation does not replace market relationships but is based on their rational utilization in the public interests. The forms of regulation that are most effective, are those which enhance rather than suppress the market signs to look for new directions for capital investments. Second, centralized control of technological progress is highly selective. Within this system the prime place belongs to the setting of priorities, indispensable for the efficacy of centralized efforts. Errors in assigning priorities lead to national economic losses whose magnitude is in proportion to the size of resources allocated for their implementation.

For this reason great importance is attached to setting priorities in national technical and economic development in all the advanced countries. In most of them priorities are chosen using mechanisms of a gradual removal of uncertainty as to the implications of alternative trends in technical development. The key elements of these mechanisms are as follows: First, institutions for long-term prognostication of technical and economic development, competition among companies making alternative products. Second, joint decision-making by experts, close contact and interaction of governmental, scientific and in-dustrial bodies in setting and implementing technological priorities continuing specification of priorities as required. Third, there is a more or less conventional set of methods for implementing the priorities in the centralized control of technological progress, including direct stimulation of companies through scientific and technical programmes, subsidies, guarantees and benefit credits, as well as methods of indirect stimulation of product development primarily through various tax and credit benefits (the most widespread being accelerated depreciation, the setting of expenditures for R&D and investments in top-priority technologies against tax). Fourth, state regulation invariably encom-passes the development of organizational, social, transport and information infrastructures. Fifth, the state takes responsibility for funding basic research, exploratory R&D in top-priority fields, as well as educational and vocational training programmes. Sixth, the state regulates the international flow of capital and technologies.

The mechanism described above could hardly be set in motion within the framework of the command economy, which is evidenced from our historical experience. Up to the most recent time allocation of resources by departmental interests depended on increase in production within an framework of the extremely stable system of economic relationships which mediated the flows of material goods among economic entities. As a result, the scope for allocation of productive resources was limited by their established growth pattern which was patently inadequate for implementing rapid technological shifts. The occasional revision of priorities in technical and economic development, necessitated by the introduction of new technologies, is accompanied by decision-making and allocation of resources to new complexes of related production. However, when the greater part of limited resources is used by a system of self-perpetuating sectors and departments, this allocation could not secure large-scale shifts in the technological structure of the economy. In spite of the seemingly powerful centralized administration, the national economy was in fact developing spon-taneously. The national economic policy was chiefly confined to imitations of technological shifts occurring in the advanced capitalist countries. Meanwhile the departmental-sectoral system of public production and science provided merely for slow and steady increase in the new-technology content of production whose large scale introduction and wide marketing encountered insurmountable departmental barriers and the handicap of the fixed sectoral allocation of resources.

In the departmental-sectoral system of economic relationships new techno-logical systems come into being along side the old ones through formation of new industries and subindustries, the latter being provided by resources for expanded production. The companies responsible for development of new technological systems, being, as a rule, incompatible with the conventional ones, do everything possible to acquire their own technological base. With the formation of new technological networks and a new technological structure the technological links between new and conventional production systems are getting weak and are breaking. The new structure continues to develop on its own. But this production through self-generated resources expands very slowly. The technological systems of the new structure have to be expanded not only through new capital but also by involving the conventional technological processes in their growth. This implies breaking the old technological networks and the corresponding economic rela-tionships, which, in turn, is impossible without dismantling or readjusting many institutions responsible for conventional technological systems. Meanwhile new technological structures are formed simultaneously with expansion of the old ones, and in the course of time several autonomous technological structures functioning in steady-state conditions of an expanded market take shape.

During the last century a succession of three technological structures could be traced in the economies of the advanced capitalist countries. The first one began in the latter quarter of the 19th century and was based on electrical power and electrical engineering. The introduction of this technological structure was accompanied by mechanization of the basic processes and corresponding changes in labor skills and education. The key structural material in industry was steel and rolled stock, coal was the chief fuel, and railways, the main conveyance. Industries within this technological structure exploited mass resources, standard machinery and equipment and unskilled labor. The development of this technological structure involved rapid urbanization and radical changes in the consumption patterns of the population and its ways of life.

Along with concentration of economic activity around the industries of the current technological structure, there appeared new, rapidly expanding spheres of public production such as the chemical industry, automotive industry and telecommunications; and new infrastructures in transport (motor roads), information transfer (telephone) and power (liquid fuels). Wide introduction of new technologies was held back by the old methods of production management, stereotyped consumption patterns and the greater part of the resources being kept within the contour of the conventional technological structure. Only the radical changes in the institutional structure which happened in the thirties, and an acute structural crisis which sharply devalued the capital invested in old technologies allowed rapid expansion of the new technological structure to start. Mass production based on Taylor–Ford methods of management as applied to new technologies, an increased demand for consumer durables arising from the new infrastructure and from the consequent stereotypes of demand made for rapid economic growth of advanced capitalist countries in the 1950s–1960s.

Our research shows that aspects of the consumption of conventional materials and the production of some of the goods characteristic of the first technological structure continued to be reproduced in the advanced capitalist countries up to the mid-sixties. Nevertheless, the key moving force in postwar technical-economic development became a new structure based on the chemical industry (including organic synthesis) and related mechanical engineering enterprises while road transport came to play the leading role. This stage was characterized by all-round mechanization and automation of many basic processes, growing specialization of production and its reorientation towards the use of quality raw materials and specialized equipment instead of mass resources and standard equipment. Electrical power engineering was developed at very fast rates and the fuel balance was shifting markedly towards oil which came to be the main source of energy. In structural materials, the use of novel materials and high quality steel increased substantially. A new type of consumer market spent heavily on services and more sophisticated durables (chiefly, household

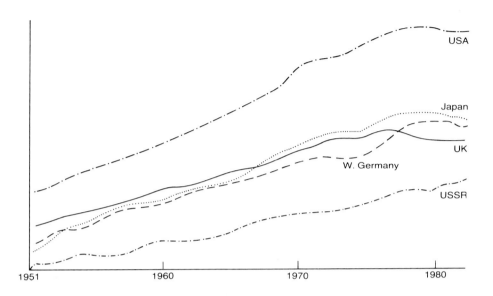

Fig. 2. General indicator of the evolution of the fourth technological structure.

electrical appliances and cars). The advent of mass secondary education raised labor skills substantially and, consequently, production standards.

By the mid-seventies the technological structure of the advanced capitalist countries achieved the limits of its expansion, which is evidenced, among other things, from the dynamics of indices pertinent to its life cycle in our analysis (Fig. 2). By this time the leading capitalist countries had stabilized the relative consumption in the basic structural materials, fuels and the consumer commodities of the current technological structure. Further technical and economic advance coincided with a new stage in technological progress.

The third technological structure, according to our schema, was based on wide-scale automation of production with microelectronics and robots, a growth of specialization of production (made possible thanks to manufacturing on industrial scale of materials with preset properties, and a new system of mass communications making wide use of computer networks and satellites.

The third technological structure in virtually all countries is currently held back by backwardness of the corresponding types of infrastructures in information (computer networks), transport (aviation), power (means of extraction, transportation and utilization of natural gas) and the social infrastructure (above all, everyday services). Other tangible barriers to a rapid advance of the new structure are insufficiently high standards and inadequate human engineering. All advanced countries are looking now for organizational and social innovations which would secure a large-scale reallocation of resources into the new technological structure. Some of these innovations, such as flexible production management, based on informal communications and creative initiative of the personnel, have gained wide currency.

The formation of the first technological structure in Russia began a bit later than in the advanced countries of the time, by the end of the 19th century. Its rapid growth was interrupted by the revolution, the Civil War that followed, and the ensuing disorganization of public production. Afterwards, during the years of industrialization of the national economy, a new impetus was given to expanding the industries of the first technological structure. This has been going on ever since.

During the last war the enforced military build-up stimulated the introduction of the new technologies of the second structure. But the latter's expansion and all-round introduction was once again interrupted, this time by the restoration of the national economy which was aimed at reviving the prewar industrial structure. Only with the second half of the fifties, after the well-known decisions were taken to emphasize the chemical industry, separate elements of the second technological structure were introduced into the civil sector of the national economy, along with the continued expansion of the first structure which still predominated in the economic sphere.

The dynamics in the consumption of conventional structural materials, production of standard metal-cutting equipment and some other indices relevant to the life cycle of the first structure show that its expansion proceeded up to the mid-seventies and is still going on. Allocation of huge national economic resources for the continual expansion of the first structure made it impossible to allocate resources to development of the second structure. As a result, the rates of introducing its basic technological systems are far behind the rates achieved by a number of advanced capitalist countries.

This conclusion is based on an analysis of the dynamic series of indices (1951–1984) which reflect the technological shifts of in the second technological structure. This work shows that in the USSR and other Eastern European countries the second technological structure is currently in its growth phase. The dynamics of indices pertaining to this structure are worthy of note: in contrast to capitalist countries the indices of Eastern European countries have the form of a straight line rather than a curve. The phase of formation here turns out to be comparatively short with the rates of growth within it being relatively high, whereas in the phase of growth of the new technological structure these rates are considerably lower than in the advanced capitalist countries. After a strong initial impulse expressed in a sharp growth of investment in the new structure and its subsequent increased contribution to the GNP, structural shifts in the socialist economies suddenly slow down.

At present the production level of the second technological structure in our economy corresponds to that of the USA and other advanced capitalist countries in the early and mid-sixties. If this tendency continues, the technological networks of the second structure would reach the limits of their expansion not earlier than 2005.

Our national economy is currently also witnessing the formation of the third technological structure. Our research points to relatively high rates of developing robotics, microprocessor-controlled machine-tools and other basic technologies. But it should be remembered that this technological structure is still in the initial phase of its life cycle. For this reason the high rates of expanding its basic systems require as yet no big resource expenditures due to their insignificant weight in the whole structure. To keep up the same rates in the future will require reallocation of resources on an ever increasing scale. This task is of exceptional complexity, which is borne out by the experience in developing the second technological structure and the manifest tendency to lag behind in the rates of introducing the most sophisticated technologies to manufacturing industry since the early eighties. Despite the powerful initial impetus in the development of the third technological structure in the USSR there is an obvious tendency to slow down. This is evidenced from some indices, which show a lagging behind in new mechanical engineering technologies and slowing rates in the output of computers. Big difficulties also arise from the incomplete development of the second structure, which is the material and technical basis for the formation of the third.

The qualitative analysis of the technological structure of the Soviet national economy and our measurements show that in fact it incorporates three technological structures going through different stages of their life cycle. Because of fundamental differences in the technical levels of their components the links between these structures are mainly of binding force, taking the form of compensatory and substitutable flows of resources. Each technological structure has its own mode of operations and its own base for growth. Their intercrossing causes real losses in the national economy owing to the fundamental differences in the technical levels of adjacent production processes.

The multistructural economy hampers progressive technological shifts. The growth and maintenance of outdated technological structures take up considerable production resources and limit the growth of a new structure. This leads to disproportions (the most obvious of which is a glut of obsolete commodities while there is an acute shortage of new), and to slowing of general rates of economic development.

It should be noted that the co-existence of several technological structures is a normal state of the economic system while the dominant structure change.

In the unfolding of structural interactions the old structure is being destroyed, which creates the prerequisites for the emergence of the new. But the growth of a new structure may proceed against the background of the continual expansion of the old. This is usual when a new structure enters into its growth phase before the life cycle of the previous one has come to a close. In this case the potentials of the new structure turn out to be limited by the continual expansion of the old one and in the absence of a vigorous structural policy a plethora arises of several coexisting and growing reproductive technological structures.

This pathological state of the multistructural economy is characterized by stratification of its growth. When two or more technological structures are each on growing on their own, structural interactions are directed not at the transformation of outdated technological structures in accordance with the needs of the new ones but at liquidation of the bottlenecks which hold back their simultaneous growth. In these conditions satisfaction of demand for primary resources becomes the most obvious bottleneck, since all the structures are based on the same raw materials. As a result, substitution effects are primarily concentrated on extractive industry, making the economic foundations too heavy from the additional demand for raw materials. There arise some other unpleasant disproportions.

In the context of the still existent old technological structure the formation of a closed contour of a new structure is impeded by the limited supply of their common resources and the missing commodities are supplied by foreign trade. The scarce high-quality resources are bought in the world market, for mass natural resources, which means an unequal exchange eventually, for the national riches are wasted to provide for the missing links in the contour of a new structure.

Another negative implication of the simultaneous existence of several structures is a stratified system of economic estimates in the growth pattern of the national economy.

In the command economy a glut of new products is not, as a rule, accompanied by falling prices, as is the case with the market economy. Neither enterprises, nor ministries, nor the State Committee for Pricing are interested in lowering prices. That's why new technology in the command economy in nearly all cases proves to be more expensive than the old technology, even if expressed per unit of the end product. Formation of each new structure involves the simultaneous establishment of new prices according to which new products are relatively more expensive. Since each technological structure grows relatively independently, while competition is non-existent and expansion of production is regulated by command assignments based priorities other than price, the differences in costs of alternative technologies in the new and old structures have virtually no effect on their general use.

Table 1. Relative price structure in the USSR and on the world market.[a]

Product	USSR prices[b]	World market prices[b]	Percentage ratio
Electrical power (1000 kWhr.)	1	1	100
Oil (1000 tns.)	1570	5000	31.4
Copper (tn.)	66	127	51.9
Steel scrap (tn.)	3.1	20	15.5
Meat (tn.)	0.26	2	13
Video recorders (pcs.)	61	18	339
Cars (pcs.)	500	250	200
TV-sets (pcs.)	75	25	300
Computers (pcs.)	8125	125	6500

[a] Source: Statistical Abstracts of the USA. Washington, 1987.

[b] Expressed in 1000 kWhr.

With the restructuring of the national economic management system along market principles pricing will become the key reference point in the behavior of economic units. Relatively high prices for new products and low prices for the long available commodities and services, including nearly all types of mineral and agricultural raw materials and products (see Table 1) will slow down progressive technological shifts in the national economy, make it impossible to reconstruct and will not build adequate foreign exchange reserves.

The above analysis warrants the conclusion that the effective integration of the Soviet economy in the world market calls for its radical reconstruction.

The scope of this reconstruction is immense. With its implementation the growth pattern of the national economy would change radically. The system of economic estimates guiding the performance of businesses would change correspondingly. But market relationships under the current structure and the respective correlation of estimates could hardly secure progressive technological shifts and reconstruction of the national economy, for the prevalence of radically new technologies indispensable for reconstruction of the national economy is disadvantageous in terms of maximizing the current profit. It should also be pointed out that in the current conditions of production concentration, with the monopoly status of many enterprises, their transition to full cost-accounting would not trigger immediate renovation of their production. Most enterprises, even in the conditions of economic reform, would be more interested in expanded growth of outdated technologies.

Moreover, reconstruction of the national economy is hardly feasible within the confines of the departmental system that still prevails in public production. The dominant position of departments in the structure of command economic relationships, to a great extent, hampers reallocation of resources from outdated industries to new. Like any large organization, an economic department seeks to expand the sphere of its activity with as few serious innovations as possible, not to disrupt the conventional communications and interrelationships.

Over the decades of departmental control over public production there have been shaped the stable contours of economic relationships which secure expanded growth and the evolutionary streamlining of the existing production-technical systems. Departmental research blocks the introduction of radical innovations. Large-scale technological shifts are carried out following decisions by the country's political leadership to set up new bureaucratic structures which later grow on their own, with no tangible effect on the growth of the rest of the national economy.

The current mechanism of technological shifts has exhausted itself, having caused a number of grave disproportions. It cannot meet the requirement to restructure the national economy which should embrace all sectors of public production rather than limit itself to evolving various new production-technical systems.

So, the circle is coming to a close. Reconstruction of the national economy and acceleration of its technical development are hardly feasible within the framework of the departmental administration of public production. On the other hand, developing market relationships are accompanied by growing chaos and take the economy away from the mapped out trajectory of reconstruction. The behavior of enterprises, irrational in national economic terms, but guided primarily by the actual economic conditions, entails stronger administrative interference. It is no accident that economic reform regularly goes astray: advances towards the market economy interact with counteroffensives by the command system, with the latter's basic institutions kept intact, for this system assimilates market institutions, bringing them down to the status of secondary appendages of the departmental-bureaucratic structures. Under the pressure of growing crises and the mounting lag in technology the economic management system has to undertake ever new cycles of transformations which, however, fail to change the situation drastically, and which leave the key problems unsolved.

To break out of this vicious circle formed by the interlocking measures taken to implement the economic reform and reconstruct the national economy, a number of radical institutional transformations are needed in the national economic management system. Below we detail a range of measures to deepen the restructuring of the socialist economic management through centralized regulation of the market which would secure a rational reconstruction of the national economy and accelerate the country's socio-economic development while preserving the basic socialist values and minimizing the already apparent social tensions.

3. How to Enhance the Economic Reform

Continuation of the current trends in restructuring the national economic management system on the principles of collective ownership will not be able to secure reconstruction of the national economy and its integration in the world market.

As it follows from the analysis made in the first part of this chapter, transformation in management of state enterprises based on quasi-collective ownership and the weakening of the conventional institutions of state regulation while the existing system of pricing and the technological structure of the national economy continue, make businesses take decisions which are not in the interests of the national economy. The contradictions inherent in the market economy between businesses and local and central governmental bodies will, under current trends in reform of public production, be manifested as inflationary processes, redistribution of profits in favor of consumption, the growing disparities in spending power among working people of different enterprises and regions, and in disruption of many conventional economic relationships. Aggravation of these contradictions may lead to disintegration of the national economic complex, disorganization of public production and uncontrolled social tensions. At present we can observe the first symptoms of mounting chaos in the consumer market, with further rationing of commodities, more frequent accidents and disasters in different spheres of public production, outbursts of strikes, violence against members of cooperatives, demands for economic autonomy of republics and regions, and aggravation of interethnic conflicts. The national economy could be turned from this patent trajectory of economic transformations only through a radical deepening of the economic reform and the speedy formation of modern market institutions.

The unreadiness of the population, including production managers, to restructure the national economic management on market principles, a lack of socio-psychological institutions and of the corresponding ethical standards, as well as the absolute necessity of reconstruction of the structure of the economy, all mean the state administration has the decisive role in implementing not only the transition to a market organization of public production but also in backing up the market institutions during the first stages of their performance. The need to reconstruct the national economy also makes it necessary to keep state responsibility for planning and regulation of economic activity, including direct methods of affecting the decisions of businesses. Certainly, the forms and methods of governmental interference in the economy will have to be essentially different.

Thus, in the state control of the national economy it is necessary to give up any attempts to plan by formal command indices. State control should solve policy-making problems involved in the long-range development of the national economy. Of course, some indices will have to be used at this time, but not as an end in themselves. The main purpose of planning is to carry out large-scale measures aimed at changing the growth pattern of the economy in conformity with social needs. These measures are not necessarily numerous, but they should give impetus in the desired direction to irreversible economic transformations. The centralized national economic management must be changed. The administrative hierarchy must be replaced by market mechanisms, and the red-tape routine in controlling economic activity by self-adjustment of independent businesses.

We have already mentioned the mounting chaos in public production under the narrowing zone of operation of the conventional institutions of national economic management, which provokes waves of counter-reform agitation. It should be added that curtailing the administrative institutions and expanding independence of state enterprises should proceed very carefully and involve new forms of integration to prevent a sharp rupture of the conventional economic relationships and to lay foundations for the varied market regulation processes. This is due to the following objective considerations. First, it is of vital importance to retain the long-standing production, scientific and technical contacts between enterprises which used to cooperate with each other. Their establishment takes, as a rule, several years, and a sharp break of these contacts through making all enterprises fully independent may disorganize production and create havoc in the national economy. Second, most enterprises are not ready today, for good reasons, for independent decision-making, lacking competent personnel, the experience and the resources for an efficient performance under the market economy. Disproportions in prices, a monopolist structure of virtually all sectors and industries, absence of market infrastructure, and economic incompetence of managers at all levels make it extremely difficult to set up market relationships by granting complete economic independence to existing enterprises. By the way, according to forecasts, nearly one third of them would prove unprofitable in the new economic conditions. This means that along with introducing market relationships it is necessary to take special measures which would compensate the market's imperfection, on the one hand, and the unpreparedness of enterprises, on the other. The currently used mechanism of public subsidies to keep unprofitable enterprises afloat, despite the market's definition of their inefficiency, runs counter to progressive technological shifts in the economy. The mechanism of market regulation should not suppress the market's signals as to a need for resource reallocation but, quite the contrary, to enhance them for timely measures to be taken to forestall the economic disintegration of enterprises. While setting up the new organization of national economic management, use should be made wherever possible of the existing structural forms (embodying the new content) to reduce to a minimum the resistance to these new structures.

In our view, the following groups of measures would promote economic reform and provide an effective mechanism for restructuring the national economy.

1. Integration of independent enterprises of similar specialization into commercial associations, to be based not on formal bureaucratic administrative machinery but on coordination of economic activity, collaborative development of long-term development strategies, and on a common center for accumulation and allocation of resources, such as a bank, a finance or investment company which is entitled to issue bonds, shares and other securities, and to purchase and sell productive assets. Interrelationships between the center and such associations should be based on market regulation mechanisms. Administration on the part of a bank should be limited to indirect general planning, tied credit and subsidies for top-priority activities, joint consultations, market transactions in the means of production, and sales and logistics services. The same centers (banks) should shoulder financial responsibility for the operations of incorporated organizations and enjoy the rights to redirect unprofitable enterprises and to sell them, and control the sale of shares and other securities.

2. Commercial associations may have the form of companies, trusts, consortia, syndicates, etc., depending on their activity and the preferences of the partners. As prototype these associations have some of the recently established intersectoral amalgamations. According to this schema, such associations should in our view be based on a corporate form of ownership, which is not only a highly effective way to set up large-scale production with its basic separation of management from ownership, but is also the foundation of today's market in the means of production which is indispensable for rational and timely reallocation of resources from outdated technologies to more sophisticated ones.

The institutions of corporate ownership along with bank credits make the basis of a capital market which is, in its turn, the foundation of a modern market economy. Lack of a capital market dictates preservation of administrative reallocation of resources and makes modern forms of marketing impossible. The independence authorized to enterprises and delegation of proprietary rights to the workforces do not guarantee higher production efficiency and could hardly serve as a reliable instrument for rational regulation of centralized reallocation of resources. As exemplified by Yugoslavia, concentration of proprietary rights in the hands of the workforce makes enterprises powerful enough in the economic sphere to exert pressure on the political leadership. In the context of the corresponding system of social institutions this engenders some insolvable contradictions manifested by rapid inflation and growing disparities among regions in their levels of development and in the living standards of the population.

To transform state enterprises into joint-stock companies needs a statutory procedure with government organizations entrusted by the state with the share capital of the transformed enterprises. Such stockholders may include the existing or newly emergent banks or holding companies founded on both governmental and non-governmental basis which are to sell out the government stocks gradually with the formation of market self-regulation institutions.

3. Simultaneously with setting up commercial associations of independent enterprises it is necessary to restructure both bodies and methods of centralized national economic management. In the new economic conditions there is no need to compile and implement comprehensive plans of the national social and economic development based on material and commodity "balances". The planning agencies should confine themselves to predicting national economic development and to providing information for decision-making at all levels of public production management. The centralized reallocation of resources should concentrate exclusively on the top-priority areas to change the growth pattern of the national economy.

To carry out these functions it would be sufficient to have a relatively small administrative apparatus in the national economy. In our view, it may be limited to a few economic ministries (industry, trade, agriculture, transport, communications, power supply and aerospace) and committees (for planning, finance, science and technology, environmental protection and social matters). The planning committee should focus on prognostic-analytic calculations rather than on balancing between formal indices and on the allocation of scarce resources. On its part, the committee for financial regulation should be deprived of any legislative power and serve merely as an instrument for reallocation of of the state budget into the areas specified, as the law requires, by the Supreme Soviet of the USSR.

4. The new organizational structure should enhance the role of banks which are eventually to become major centers of national economic management and reallocation of resources. It is advisable to set up a two-level financial and credit system. The first level is to incorporate governmental institutions which regulate financial flows, control the quality of money in circulation and secure the stability of the financial system. The second level is made up of networks of independent commercial banks, including the above-mentioned financial centers of commercial associations of enterprises. The banks of the second level are to be the basis of the capital market and to become centers of accumulation and reallocation of resources. Responding to state guarantees and support, they will become reliable supporters of centralized regulation of business activity.

It should be pointed out that implementation of the above proposals is not so intricate as it may seem at first sight. At any rate this approach is easier and more effective than one based on granting independence to enterprises and on marketing which would encounter the unbreakable resistance of both the central managerial bodies in the national economy and the enterprises themselves.

Commercial associations of enterprises could be based on the existing structures such as sectoral ministries and departments and state industrial amalgamations which will have, of course, to be radically readjusted and regrouped in accordance with their preferences and considerations of national economic effectiveness.

In their turn, economic ministries and committees of a new type could be formed on the basis of the corresponding bureaus of the USSR Council of Ministers employing the best experts from the staff of the existing ministries and committees, which are to be dissolved according to our schema.

It would be worthwhile to enrol specialists from the central economic departments in new institutions of economic regulation, such as investment companies, banks, stock and commodity exchanges, etc.

Dismantling of the institutions of the command economy should be accompanied by the creation of new institutions of state regulation of public production readjusted on the principles of the market economy. The point here is not to give up the state centralized administration of the national economy but to change its forms.

Centralized planning should concentrate exclusively on solving fundamental, strategic problems of long-term economic development which cannot be solved by the market self-adjustment of economic entities. Such planning involves, first, a social element with a system of social guarantees, levelling up the living conditions of people in different regions of the country, with a social security system, protection of the population against inflation, and the like. Second, ecological protection, including environmental protection, introduction of ecologically safe technologies, etc. Third, a unified policy for science and technology covering top-priority areas and developing a mechanism for replacing outdated technologies by new, updating national production potentials, encouragement of inventors and other innovative activities, keeping up of the appropriate scientific standards, training and retraining of personnel, etc. Fourth, the economic strategy and tactics to embrace the unified taxation system, finance and credit policy and economic control, money circulation and pricing, foreign-economic relations, etc. Fifth, the foreign policy and defence initiative.

All the remaining problems involved in the current economic changes, such as defining the economic activity of associations and enterprises, production and sales, consumer services, forms and methods of production management and payments should be solved by companies and local government. Besides, for their resolution use could be made of the mechanisms of the regulated market based on direct and indirect methods of the state, such as awarding government orders on the competitive-contractual basis with the corresponding system of financial and logistic backing; benefit taxes; tied benefit credits; by a partial covering of period and project costs of enterprises, etc.

In accordance with rationalization of managerial functions the techniques of planning should also be specified. First, it is necessary to give up the very idea of the so called balanced plan which is still the key link in the existing administrative-command system. In the new economic conditions balances between production and consumption will be regulated by the market.

Plans of national social and economic development should be based on programmes aimed at solving definite problems. Of course, implementation of each programme would involve balancing the resources required with those available. But such balancing depends on how the programme is to be implemented and has nothing to do with the national economic balances calculated from formal indices of production and consumption of material resources.

Second, the role of planning standards should be changed radically. With the rapid advance of technology (in some industries a new generation of technology is introduced every two or three years), long-term stable standards for consumption of resources inevitably hamper technological progress. A goal-directed plan does not need stable standards. Though to work out a programme one must calculate the cost of resources and refer to the respective standards, these are not to be applied as a parameter of management but to a quite different purpose, which makes less strict demands as to their authenticity.

Third, the central agencies of state administration should orient themselves to solving long-range problems in national socio-economic development, in other words, to selecting, quantifying and implementing top-priority guidelines for national economic development. Their purpose is to set in motion the search for and appraisal of alternatives for national long-range economic development and to work out programmes to realize the chosen priorities, rather than to supervise a formally balanced development of the national economic complex and to shape general economic, intersectoral and regional proportions. It should be understood that because of their inherent uncertainty these proportions cannot be envisaged for a lengthy period of time. Such planning only impedes economic development and makes our lagging behind the advanced Western countries chronic. Along with the central national economic agencies focusing their efforts on policy relating to implementing the top-priority tasks, the use of consultancy services and experts, including those invited from outside, should be widely encouraged.

With programmes becoming the basic object of national economic planning, the whole procedure of compiling and approving plans for national social and economic development changes drastically. Each programme should be separately considered, appraised and approved by the supreme legislative body, which makes it useless to draw comprehensive master plans for national social and economic development, as well as to compile complete and regularly revised listings of national scientific and technical programmes.

A goal-directed plan lends extra substance to government orders which should serve as a kind of bridge between social consumption and new technologies.

The renouncing of formal plans for "everything", a transition to goal-directed methods and the priority principle in central regulation of technical and economic development implies a readjustment of management techniques and of the methods of economic agencies, which should be transformed from being the centers of economic authority into decision-making centers whose prime function is to act as an expeditor of the joint decisions made by experts.

The restructuring and reorientation of the centralized national economic management system, with the balances of current production and consumption geared at solving prospective problems of national long-range technical and economic development, should include the establishment of a subsystem to control introduction of new technologies, research and development and selection of the most efficient innovations out of as many initial proposals as possible. The high degree of uncertainty of innovation processes makes centralized methods of their control hardly feasible. In R&D work and in the transfer of its results into production it is extremely important to secure the cooperation of specialists, with the parties in the innovation process assisting in its successful implementation, and horizontal communications. And centralized administration with its striving to limit the diversity of objects to be controlled proves ruinous for innovation processes. Only a great diversity of initial proposals can guarantee success in the conditions of uncertainty.

At the same time, there are far more R&D designs than there are potential practical applications and this necessitates selection of the most effective. The criteria for selection in a planned economy cannot be formed spontaneously, but have to be regulated from the center to meet the targets of national economic development. The growing expenses for R&D which now exceed 5% of the national income, and high costs of modern R&D dictate a need for centralized regulation of expenditure in the innovation sector. The central bodies are also responsible for outlining the R&D guidelines in conformity with national economic priorities.

So, control over the innovation sector should be based on spontaneous processes with centralized regulation of their direction. Among the best such processes are the following: flexible organic structures for controlling enterprises and research institutes, goal programming, R&D on a competitive and contractual basis and stepwise promotion of projects, as well as a small research intensive company as the key participant in the innovation sector. This sector should function in such a way as to promote as many diverse projects as possible at the initial stage, then to cut the number of alternative projects as they go from the R&D stage to prototype development and, finally, to let only the most sophisticated technologies reach the stage of serial and mass production. The part played by spontaneous processes should thereby diminish whereas the role of centralized regulation is increasing with the progression of projects from the R&D stage to mass production.

There is usually a high uncertainty at the early stages of design as to the potential value of innovations, and there is hardly anyone competent to see the promise they hold, except for the design engineers themselves. Therefore scientific and technical researchers should be given as much free hand as possible in their R&D work. This could be provided by a small independent research intensive company, subsidized within the framework of the corresponding scientific and technical programme. Such a company should be fully independent in its operation, while its growth depends directly on its progress. The right to set up such companies at government expense should be accorded to the designers of the most promising projects through the mechanism of appropriate contests. It is worth encouraging R&D cooperatives as much as possible.

Since the number of projects, as a rule, exceeds by far the number whose potential will one day be implemented, the outlooks seem uncertain in each case, while the costs rapidly increase with each successive stage in the research–production cycle, the only possible way to remove uncertainty and select the best projects to be financed by the government is a stepwise progression of projects and their expert examination before each subsequent stage. The projects deemed promising one stage are cleared for the next stage of financing. The rigidity of selection varies, depending on the significance for the national economy of the problems involved and the volume of resources allocated to a given programme. In actual fact, ninety percent or more of projects would be rejected at some stage of the applied research process.

Beginning at the stage of pilot production, the organizations which introduce and promote scientific and technical projects should be financed through long-term credits under the control of a creditor, and it would be advisable to establish a special bank for scientific and technical progress.

A characteristic feature of the new technological structure is integration of science, production and education. But at present formation of an integrated scientific-educational-industrial complex is blocked by departmental barriers which divide the Academy of Sciences from the schools of higher learning and from industry.

Integration of science, higher education and industry involves a reorganization of R&D and higher education. What is needed is to make academic institutes and universities fully independent. Centralized regulation of R&D and the higher education system should become oriented towards goals dictated by the priorities of national development and given flexibility, primarily through indirect methods of control. These principles call for a radical adjustment of the Academy of Sciences, the State Committee for Science and Technology and the Committee for Public Education. Their functions are to be limited to the distribution of state money to science and education, the establishment and provision of the information, material-technical and, partly, organizational infrastructures of scientific research and educational processes, and prediction of long-range directions in R&D and the demands of the national economy for specialist skills.

The revision of the functions of the Academy of Sciences and the State Committee for Science and Technology can be started on the basis of the existing structures of these institutions. With this in view, their staff and the associated scientific and interdepartmental councils should reorient their activities to identifying the long-range directions in R&D, compiling research programmes and securing their realization through distribution of the budgetary allocations and those from the other interested parties on a competitive-contractual basis. Major aspects of their future work will be establishment of cost-accounting centers for contractual research, technical and engineering centers, readjustment of the system of scientific and technical information to use the most sophisticated means of data transfer, processing and storage, providing services in logistics and research work, and tied subsidizing of research institutions. There will be no need in this case to draw formal R&D plans and proceed with many other labor-consuming formalities. The cumbersome departmental control over scientific institutions will be replaced by a far more efficient system of non-departmental expert examination of specific research projects.

The administrative system in higher education can be adjusted in a similar way. Compilation of methods, aids and curricula and many other matters involved in the educational process will be settled by higher educational establishments themselves, while the Committee for Public Education concentrates on tied distribution of budgetary allocations for higher education.

In accordance with our schemata for industrial adjustment, research institutes, design bureaus and other research institutions can be incorporated in the newly formed groups of enterprises or become independent. In any case most of them should be freed from departmental control. The role in R&D regulation of the remaining ministries, in conformity with the general principles, should be reduced to distribution of budgetary and outside allocations for contractual R&D and specialist training and to establishing the information, organizational and logistic infrastructures.

Alongside the adjustment of the existing institutes, new ones must come into being, primarily, non-departmental, fully independent research and educational establishments, such as consortia, centers, institutes, laboratories, cooperatives, etc. to be funded by interested enterprises and organizations. A major task is stimulation of R&D ventures, which calls for a corresponding infrastructure of venture financing institutions. It is expedient to set up special foundations at the State Committee for Science and Technology and the Academy of Sciences, which could act as guarantors of credits granted to venture projects. Another noteworthy idea is to set up innovation banks to finance promising R&D ventures.

It is also advisable to institutionalize the enhanced role of science in the life of society and to establish, for instance, a scientific council under the Chairman of the USSR Supreme Soviet, with appropriate posts in the staff of the Chairman of the Council of Ministers, and commissions on scientific and technical policy at the Supreme Soviet of the USSR.

A workable mechanism of personnel selection is a major prerequisite of efficient scientific research. Virtually all national research institutes have on their staff very many enterprising people who are, however, not qualified to do scientific work. New, highly profitable forms of economic activity would mean such people no longer inhibit scientific advance. But to normalize the psychological atmosphere of research work we need to get rid of the red tape in science organization, diversify the forms of scientific activity and promote the fundamental principles of scientific ethics.

The restructuring of the national economic management system along the lines specified above is far from achieving a complete solution of the problems involved in reconstruction of the national economy in the context of the emergent market, rather it creates the conditions for their solution. The main problems facing the Soviet economy, as was pointed out above, are of a structural nature and will take a lot of time to be solved under the regulated market economy.

One of the most serious barriers to reconstruction of the national economy is the outdated vocational training system, which corresponds to the conventional multiform technological structure of public production. With the latter being radically restructured, vocational skills of many millions of people will be rapidly devalued, which would inevitably cause them to resist the unfolding changes. Besides, the relief of the national economy from the burden of redundant economic activity will entail a fall in production volumes, with lower real incomes of the employed and resultant social tensions. The difficulties involved in the economic restructuring will be further aggravated by the uneven load carried by different regions of the country. The traditional industrial zones with their vast, hopelessly obsolete industries, such as the Urals, the Volga area, and the south of the Ukraine, may find themselves in a disastrous state. To prevent this, special compensatory measures will have to be taken to disperse the inevitable structural unemployment, retrain the personnel and introduce new technologies. It would also be necessary to provide legal grounds for mass migration of the population, including the abolition of the existing system of domicile registration and the lease out of state housing in addition to private and cooperative house-building.

A normal market of labor and the restructuring of the national economy will also need an efficient system of personnel training and retraining; liquidation of the rigid staff classifications; the abolition of groundless job demarcation and the wage ceiling; a reform in the labor legislation to facilitate sacking of redundant personnel, and a reform in the social security system to ensure acceptable living standards for all categories of the population. A tax reform should provide for a socially fair distribution of incomes and alleviate the negative aftermaths of their growing differentiation. It would also be advisable to transform trade unions to make them an effective pressure group for the working people's rights, primarily, with regard to raising their wages.

The centralized efforts should be directed to readjustment of the public education system and personnel retraining. This is of vital importance since the bottleneck in developing new technologies lies not in the shortage of material resources but in an urgent need of personnel with the required skills and knowledge. At present the production of specialists with a higher education in the USSR corresponds to the technological requirements of half a century back. Under the multiform structure of the national economy, transition of higher educational institutes to cost-accounting principles will not be enough to ensure their rapid readjustment to meet the current requirements of the national economy. Centralized efforts are needed to expand the training of specialists in information technologies, along with making the higher educational institutes more independent in the choice of curricula, courses and methods of teaching.

A similar radical readjustment is overdue in secondary education whose teaching and pastoral methods go back to the years of the Stalinist dictatorship. Both the subjects to be taught and the teaching methods should be revised. The methods of teaching must become democratic, with development of self-management, to raise a generation of thinking people with high standards of civic duty. School-leavers must be equipped with knowledge of computers and the basics of information technologies.

Of course, the readjustment of public education should begin, in the first place, with radical changes in the teacher training system.

The policy towards location of industries should also be drastically revised. Industries of the new technology are characterized by a low consumption of resources, smaller production runs, high demands for a pure environment, heavy dependence on science and a need for highly skilled personnel. For their rapid development availability of information and social infrastructures, proximity to educated and highly qualified consumers have greater significance than proximity to suppliers, sources of raw materials, etc. As evidenced from the experience of the countries advanced in the new technological structure, these new industries concentrate, as a rule, in close proximity to research and educational centers. They represent kinds of "growth poles" which secure rapid technical and economic development of the entire country without claiming too much in the way of natural resources and freight traffic.

Such patterns for the location of the new technological industries reveal the fallacy of the still popular large-scale programmes for developing the expanses of Siberia, Far East and Central Asia. Vast resources invested in these regions of unfavorable climatic and social conditions cannot bring noticeable returns at the current stage of technical and economic development. Quite the reverse, they help to maintain the structural disproportions, to say nothing of the irreplaceable losses through damage to the environment and degradation of the population living in such extreme conditions.

Just look at the location of new industries in the countries which have scored signal successes in developing modern microelectronic technologies, such as Japan, Taiwan, South Korea and the USA, to see that such industries do not need vast territories, no matter how rich in natural resources. On the contrary, they are concentrated in the inhabited regions with a developed social infrastructure, skilled labor and thriving markets. In our conditions industries of the new technological structure should also be located primarily in the highly developed areas of the European part of the country, replacing the hopelessly outdated and ecologically harmful industries which burden in the national economy. The Ukraine, central regions of the European part of the Russian Federation, Byelorussia, the Volga area, and the Urals are among the regions which have good outlooks for concentrated investment in new industries. Such policy of location of new technologies will simultaneously solve the problem of reconstructing the old industrial regions with minimal social and economic costs, and will revive the small towns of the Russian non-chernozem zone.

There is also promise for the accelerated development of new technologies in the southern parts of the Far-Eastern region – in the Primorski and the Khabarovsk Territories and on Kamchatka. In addition to a sufficiently developed social infrastructure, a favorable climate and skilled manpower resources, this region has two other major advantages. First,the ecological situation here is relatively safe, which is imperative for location of some basic technologies of the new structure, among them the production of integrated circuits and electronic components. Second, it is in close proximity to a new, rapidly developing center of the world economy. According to numerous forecasts, Japan is to become the leader of the forthcoming stage in technical and economic progress, while the Asiatic part of the Pacific region is currently taking shape as a major economic zone in the new international division of labor. The import of technologies, know-how and human engineering from Japan, as well as from Korea, Singapore and Taiwan, will give powerful impetus to long-range technical development of the national economy in the next fifty years. The geographical proximity of the southern part of the Far-Eastern region to the future world center of technical and economic progress is its major advantage in production development of the latest technological structure. A key role in the formation of the new technological structure should be assigned to technical and engineering centers to be set up in localities of the country well endowed with educational facilities. Another promising idea is to found a number of science parks where scientific, educational and production potentials would create the "critical mass" required for cascade-like introduction of new technologies. We can transform into such science parks Tver, Dnepropetrovsk, Pereslavl, Minsk, Ulyanovsk, as well as some satellite towns of Moscow, Leningrad, Kiev, Sverdlovsk, Novosibirsk, Khabarovsk and Vladivostok. A key element of the new strategy in location of the industry is the establishment of modern information and transport infrastructures. Indispensable to the large-scale introduction of new information technologies is establishment of nation-wide computer networks and databases. The logistics system and freight transport must be made more flexible, reliable and rapid through emphasizing the development of aviation and pipelines, and integrating transport systems. The readjustment of the national economy should be oriented today to integrating the country into the world economic system and changing its place in the international division of labor. The top-priority tasks in a long-range development of the national economy should include winning of advanced positions in world markets by science-based production with a high share of added value. At the same time we should curtail and gradually take outside the country ecologically harmful industries which consume excessive power and natural resources to produce goods and commodities with a low added value, passing to a policy of conservation of the natural resources. The changing of the foreign trade balance structure with a view to turning from the export of raw materials to the export primarily of sophisticated products is among the most important targets of a long-range national economic policy. We should also strive to change the import structure to stimulate the import of technologies and key products of the new technological structure and to restrain, wherever possible, the mass import of finished products in the priority spheres of long-range economic development. At the same time the order of the day is a wider import of finished products for the non-priority spheres of the national economy whose production could hardly become advantageous in the international division of labor. This concerns primarily those traditional branches of the second technological structure which are currently in the phase of maturity and decline, having exhausted the potential of their growth in the world economy where Soviet industry has failed to produce competitive output. These branches include the organic synthesis industry, the agricultural, chemicals and machinery industries, automotive industry, machinery for food and light industries. The import of finished products is acceptable also in the markets of the new technological structure in consonance with the unfolding international cooperation in relation to new technologies.

The effectiveness of foreign trade is determined by its organization. The state monopoly of foreign trade in the current conditions of diversified markets and the administrative-departmental system of national economic management have brought about a monopoly of separate departments, large-scale corruption and enormous national economic losses. The fact that state enterprises and cooperatives are now entitled to break into the international market is likely to normalize foreign trade. On the other hand, the definite complexities of this activity and an acute shortage of experts in the field make it more expedient to set up specialized foreign-trade companies in various industries to provide the appropriate services and consultations for enterprises. The proposals offered above on industrial readjustment may be extended by recommending establishment of trade-broker firms within the industrial associations to look after the sales of finished products and the deliveries of raw materials and semi-finished products in the foreign and domestic markets. It goes without saying that these firms are not to be entitled to any legalized monopoly of the foreign trade but will have to prove the right to their existence by the quality of their services.

To undermine the monopoly of large enterprises and to stimulate competition in the domestic market, foreign companies should be given an opportunity to sell their products on the territory of the USSR, both through Soviet and joint trade firms, and on their own. In this way the state monopoly of foreign trade is to be replaced by its state regulation by means of a flexible tariff policy and other instruments of import control, such as quotas, import quality control, currency exchange rates, taxes on foreign-trade profits, etc. The state must also reserve the right to reallocation of currency resources with the following levers to be used for this purpose: collection of taxes on foreign-trade profits in the corresponding currency, keeping foreign currency exclusively in a state bank, as well as forced currency exchange on strictly statutory terms. The state must also retain complete control over the export of capital to prevent its drain from the priority sectors of the national economy. The export of commodities should also be regulated by the state. Intensive foreign-trade activity should, as a rule, emerge only after a glut of the corresponding goods on the home market. At the same time, development of industries which are of priority importance for the national economy must be protected by high tariff barriers until they become competitive in the world market. The state control should extend over the import of foreign capital to protect the developing priority industries in the initial phases of their life cycle and to keep the foreign-trade balance within tolerable limits. At the same time the state should encourage the import of new technologies and know-how by obtaining licenses, setting up joint ventures, specialist exchanges, etc. It is advisable to grant public subsidies to import licensees in the priority directions of technical and economic development and to provide access to the foreign market for technologies of all national enterprises. To remove the monopoly status of separate enterprises in exploitation of foreign technology, the state must encourage the import of similar technologies by their competitors, as well as propagate advanced foreign know-how.

Liberalization of foreign economic activity should proceed gradually and be accompanied by the corresponding legal and economic securities. During the first stage it is necessary to achieve domestic convertibility of the rouble (of crucial importance in this respect is the forthcoming price reform), to facilitate the establishment of joint ventures (including those with a greater percentage of foreign capital) and to expand the foreign trade of national enterprises and trade-broker firms, to adopt laws on licence relationships inside the country, and to readjust the activity of the national economic management bodies to make them take systematic account of the foreign-economic networks involved in technical and economic development. To solve the latter task as it arises in the transition to the new system of public production management, it is advisable to merge the Ministries of Industry and of Foreign Trade to commit the latter's activity to the long-range targets of the national economic policy. At the same time free trade zones are to be set up in the Baltic republics, in the south of the Ukraine, the Far East (in the Primorski and, partly, in the Khabarovsk Territories) and in Armenia – to accumulate practical experience in international economic cooperation, to change the attitude of other countries to the Soviet market and expand the import of foreign technologies.

For the second stage we propose considerable expansion of the operations of Soviet enterprises in the foreign market, including the import of capital under state control, to authorize direct sales of imported commodities by foreign and Soviet companies in the domestic market, and to secure the convertibility of the rouble. During the same stage a number of laws acts should be enacted to regulate the foreign economic activity of Soviet enterprises as well as the operations of foreign enterprises in the Soviet market. The appropriate conditions are to be created to make it possible for the USSR to join international trade and financial organizations. During the third stage it would be possible to liberalize the principles of foreign economic relations, including the export and import of capital, to make the rouble freely convertible and secure an effective state control through a wide range of methods for indirect regulation of foreign economic activity.

The dates of gradual liberalization of the foreign trade are difficult to predict, for they will depend on the economic and political transformations inside the country, as well as on the rates of technical development of public production and on how soon Soviet products become competitive in the world market.

One of the major conditions for raising the technical levels of public production and for removing the structural disproportions in the national economy is conversion from military production with a transfer of advanced technologies from the military-industrial complex. The latter's exclusive, top-priority status in allocation of resources for many years entailed a growing technical lag of civil industries and was one of the main reasons for the technological multiformity of the national economy. The latest technological industries are currently concentrated almost exclusively in the military-industrial complex. Being geared to military production, they make virtually no contribution either to social needs or to public production. The transfer of technologies from military to civil industries, owing to an essential difference in their priorities and a lack of economic incentives is currently reduced primarily to handing over the spoilage rejected by the brass! This adds to the centrifugal tendencies between military and civil mechanical engineering, to the latter's striving to develop everything on its own, despite the availability of the know-how of sophisticated technologies. The priority status of the military-industrial complex in the command economy was clearly worsening the country's technological lag. Concentration of currency resources in this complex caused inefficient utilization of the imported technologies, to say nothing of the embargo on the supplies of new technologies and commodities to the Soviet Union. A key requisite for efficient utilization of imported technologies is their commercial development, including their adaptation to the local market, and their streamlining and application through national R&D in the corresponding field. When imported technologies are focused in the military-industrial complex, development proceeds towards ends which have nothing to do with either raising economic efficiency, satisfying social needs or enhancing the competitiveness of Soviet products. As a result, instead of acting to close the technology gap the import of progressive technologies has no positive effect on technical development of industry. Concentration of high-quality resources, primarily of key personnel in the military-industrial complex paralyses technical development of all other industries, which eventually causes difficulties of supply for the military-industrial complex itself, and its further "swelling", bringing the threat of collapse of the whole economic system. The currently practiced placing of orders for consumer goods with military industries would not solve the problem. What we need today is simply a transfer of sophisticated technologies from the military-industrial complex, their conversion to secure economic efficiency and their wide application. To achieve this, the military-industrial complex must be radically restructured along the principles offered above for industrial readjustment.

There are two possible pathways. First, the association of enterprises can incorporate military enterprises which are in line with the specialization of the whole group or correspond to the structure of its production requirements. Second, it is feasible, without special preferences, to bring together enterprises specializing in some definite types of military production, to diversify their economic activity and include in these groups the corresponding civil enterprises and trade firms. In both cases military enterprises must be self-dependent and transferred to full cost-accounting, with the respective departments disbanded, while control over the military-industrial complex must be exercised through orders placed by the Ministry of Defence and distributed on the competitive basis. Defence allocations and their proper spending must be subject to close public control through the Supreme Soviet of the USSR and its commissions.

Readjustment of the military-industrial complex and a transfer of technologies from the defence to civil industries call for a number of political actions. The most important of them are as follows. First, the priority of military production and the quantity of defence spending should be subject to regular expert examination in public and to control by the public governing bodies. Second, it is necessary (and also indispensable for realizing the first point) to do away with the extreme secrecy of the military-industrial complex, which not only has long lost its relevance to defence in view of the present-day capabilities of space reconnaissance and the recent initiatives aimed at building confidence among nations and among the opposing military alliances but has turned into a real brake on technological progress. It is most likely that its sole aim today is to conceal the ineffectiveness of military production. Maintenance of the conventional system of secrets not only paralyses any progress in the conversion field but blocks reconstruction of the whole national economy to meet the demands of the new technological structure. Its component industries are not only of "defence importance" though currently concentrated in the military-industrial complex, they need a good information infrastructure for a wide exchange of R&D data and results among various organizations and specialists not only in different regions but also in different countries. Third, the import of technologies and development of innovations for military purposes should be accompanied by a search for their commercial uses and their transfer to civil industries.

With time, the gap between the military-industrial complex and the other spheres of public production should be gradually overcome. A transition to information computer technologies implicit in the new structure involves radical changes in the conventional methods of human engineering, and the style and standards of production management. The world experience in production development within the new structure shows that introduction of new technologies implies new methods in management engineering. The traditional Taylor–Ford principles of human engineering have proved inadequate to the new technological conditions. Simultaneously with an all-round automation of production the companies that have prospered have been those that adopted the new human engineering system, indispensable for efficient utilization of new technologies. The characteristic features of this system are the giving up of traditional linear-functional structures of management based on the rigid hierarchy of rights and duties and the transition to flexible and dynamic goal-oriented structures without a strict hierarchy and with a developed system of horizontal informal relationships; the giving up of official instructions and regulations and narrow specialization of personnel and the transition to the system of continual rotation and retraining of personnel. Another feature of the new human engineering system is a transition from cumbersome formal routine control to informal self-control of work teams based on trust in co-workers and joint decision-making. The guiding principle in manpower management becomes not bureaucratic control and providing material incentives, but cultivation of an adequate system of values in personnel.

The new principles of human engineering correspond to the person acquiring a new status in production which requires not a mechanical execution of primitive operations but a deep understanding of technological processes and conscious initiative in their improvement. To control an intricate, constantly changing production process it becomes less important to have the mechanical skills for routine operations, than to be able to define and solve problems relating to its improvement, to perceive and accept innovations, and to stay in informal daily contacts. The production process in the current conditions of technological progress is undergoing continual changes, while its management system also has to be constantly updated in the corresponding way.

Introduction of the new methods of production management, which differ drastically from the customary red tape, is a process of exceptional complexity. It not only cuts across outdated stereotypes in the thinking of executives and a low expectation of the managerial staff, but across the generally extremely low standards of human relationships in production. The long-standing proliferation of administrative structures has made many people indifferent to their work, and to overcome this attitude through the latest management engineering methods is one of the requisites for a rapid formation of the new technological structure in the national economy. Public organizations, governmental bodies and a network of consultancy firms, whose establishment is one of our immediate tasks, may prove instrumental in this respect. The latter is only part of the more general task of shaping the new ethics of economic activity. Enterprise, an efficient and bona fide management, initiative, innovations, and prompt and effective satisfaction of social needs should be rewarded not only in material terms. These qualities should be elevated to the status of high moral values through their intensive cultivation by the mass media by drawing the new positive image of a contemporary hero, an enterprising, honest, inventive and prosperous innovator.

Perhaps, this is the most complicated task in introducing the market relationships into the Soviet economy. To change the public attitude to business enterprise, this is the vital condition for stable and efficient marketing. The suggestions offered above on how to dismantle old and establish new institutions to regulate public production may shape only the forms, whereas their magnitude and content would depend on the will of the people to give them substance.

Chapter V
Problems of transition to new forms of management

B.Z. Milner

Professor of Economics, Acting Director of the Institute of Economics, USSR Academy of Sciences

Restructuring of economic management has entered the decisive and at the same time the critical stage. Old structures that were inherited from the administrative and commanding system are already incapable of "mating" at any level with the new economic mechanism, while the new forms of management have not yet gathered momentum and are emerging with great difficulties, encountering bitter resistance from the bureaucratic apparatus. Quite often, something is introduced which is new only in form, not in content. The vices of the old structures are perceptible through the new appearance; still alive is the persistent aspiration to usurp all the rights of the top level of sectoral management and concentrate them all in the lower units. It is under these circumstances that spontaneous renewal is under way, now in one place, now in another, of the administrative and command system which has been and still remains the major obstacle to a radical reform of economic management.

There is no alternative to new, up-to-date forms of management which would create perfect, realistic conditions for the introduction of cost-accounting at enterprises, and for all-round application of economic methods. This problem must be solved in different ways depending on where in each sector productive activity is concentrated, on the geographic region, on the character of technological ties and co-operation, and on the opportunities for direct interaction with consumers.

Utilization of the potentialities of the diversifying of the ways in which property can be held is of paramount significance for the consolidation of the organizational forms of management that are adequate for the new economic mechanism. One cannot fail to see that a major progress of the economic theory has consisted lately in the establishment of a direct link between the forms in which socialist property is held and the forms of management. Reallocation to such ownership as All-Union, republican, municipal or communal, local (which may be publicly or cooperatively owned property) has already proved to be fruitful and practicable. Consistent and purposeful establishment of various forms of cooperative and individual property ownership offers new

opportunities of effectiveness. Widely known is the great potential of the switch-
over of labor collectives to contracting and leasing within the framework of the
state, as well as that of cooperative and municipal forms of property. Joint-stock
property is increasingly forcing its way through.

Rationally developed pluralism in the forms of property ownership provides
a solution for many imminent problems facing the restructuring of economic
management. Such problems include separation of the functions of the state
economic management from the functions of direct economic management;
diversity of forms of the basic productive unit; transition from chiefly vertical
to mainly horizontal structures and ties; development of associative forms of
economic management. Besides, diversity of the forms of property ownership
requires the application of various forms of self-management, and the specific
combination of independence and responsibility of a labor collective and an
individual worker, lower and upper links in the system of economic management.
It is important to make use of various means – economic and organizational –
so that every labor collective, and every worker can become a master.

The same main process is now taking place in the agrarian sector. The need
for the diversified forms of economic management, a flexible combination of
different forms of property ownership, extensive democratization of production
relations and self-management are currently the leading principles in building
up systems of production management. It is considered expedient (and was
unequivocally stated in the decisions of the March, 1989 Plenum of the
CPSU Central Committee) that districts, regions, territories and autonomous
republics should have councils, unions, associations and societies elected by
labor collectives, to provide assistance to industry on cooperative principles.

Practical experience suggested an essentially new approach to the organizational structures, with the paths of management decisions leading from
"bottom to top" – from the basic link, i.e. the enterprise – rather than from "top
to bottom". Hence, emphasis is shifted from hierarchical vertical structures to
horizontal ties of many different kinds, and to integration processes in the lower
productive and economic systems which in their turn depend upon the growing
socialization of production. All this exerts a genuinely revolutionizing influence
upon the emergence of such new organizational forms of economic activity
as socialist concerns, inter-sectoral associations, unions and other voluntary
organizations of enterprises which in the best possible way suit their diverse
interests. It is a free, rather than directed or enforced combination of factors
of intensive economic management that encourages significant developments
of productive forces. It is also obvious, that this gave rise to a very strong trend
towards lifting the state control over the appropriation processes, and the forms
of economic management and administration.

Consolidation and development of new organizational forms of economic management – socialist concerns, inter-sectoral associations, voluntary organizations, etc. – makes it possible drastically to change the functions of the ministries and steadily reduce their number. In these conditions, as is very well put in a resolution of the Congress of the USSR People's Deputies, "the role of the Center should consist in the creation of economic and legal conditions for effective economic activities at all levels, in the development of the All-Union infrastructure, in the implementation of national scientific and technical, financial and taxation policy, in ensuring social protection for citizens."

Building-up of large production and economic complexes reflects continuous processes of socialization and concentration of production and therefore meets the demand for separation of economic life from the state, with decentralization of the production management. Some of the large structures that are emerging comprise all stages of technological processes – from procurement of raw materials to release of final product. Others mainly concentrate their efforts on one specific stage (extraction or processing of raw materials), or are engaged in mass production of separate parts of the product. There are also those that cover huge volumes of services. Many different products in small quantities, or one product on a huge scale. Despite a large number of variants, the main idea remains unchanged: concentration of resources, capacities, production facilities of different types meets mass demand, accelerates scientific and technological progress, and reduces costs.

But negative aspects of large-scale organizations should also be taken into account in the formation of big concerns and inter-sectoral associations. History teaches that large structures have a distinct trend towards linear growth, bureaucratic procedures, proliferation of administration, drawing in of greater and greater numbers of people and volumes of material and financial resources.

These trends can be overcome through decentralization of decision-making, application of the mechanism of prices and incentives, placing of economic responsibility on each manufacturer for the results of his work. An important role will be assigned to further development of the socialist market as a means of overcoming the hypertrophy of vertical links in the economy.

However, there is another contradiction to be resolved. Weakening and elimination of the monopoly of ministries and central economic departments will produce the monopoly of major cartels, corporations and other mass producers. In this case, the harmful effects of concentration of production and management must be opposed by the formation of parallel structures with similar specialization, such as consumer associations who possess all the necessary rights to exercise control over the nature and price of the product and to exert influence upon manufacturers through a system of economic and legal measures.

Treating "megalomania" as a vice, an economically unjustified tendency, and making a comparative analysis of dimensions of various production systems (e.g., Soviet and foreign ones), it is necessary in each particular case to reduce the systems to a comparable form and to use the same criteria in their assessment. And above all, care should be taken not to confuse large economic structures (corporations, companies, conglomerations, consortia, etc.) with individual enterprises, big plants. For instance, ZIL (Moscow Automobile Plant named after Likhachev) is not comparable with General Motors, or Electrosila Plant with General Electric, and no oil refinery is comparable with Shell. The difference in the volume of production, in the number of employees and in the mass of profit is not less than 8 to 10-fold. When speaking about "megalomania", we have in mind expansion of individually selected plants which we call enterprises. However, economic structures comprising whole groups of large, medium and small enterprises and specialized cost-accounting functional centers are something different. These are organizational and economic mechanisms representing effective combination of resources.

Today our socialist companies are to be based on a management system free from ministries, an escape from departmental dictates and an expansion of the role of cost-accounting, rather than a direct analogue of Western forms of production management. Direct transfer of management techniques is, of course, out of the question. Its brevity and connotations of industrial might make the term "company" popular; besides, it is easier to pronounce than, say, "inter-sectoral state association", "All-Union cost-accounting production association", etc. But in any case, it is incentives, rather than orders, that must become "the management language" of socialist companies: economic but not administrative interests, partnership but not subordination.

The fundamental difference between socialist companies and the present sectoral management structure is that the former lie entirely within the cost-accounting zone, in the sphere of self-repayment and self-finance. Blending of the ministerial apparatus with subordinate enterprises has always made it impossible to discern where "superior" management ends and real cost-accounting begins, who depends on whom and who influences what, where the boss is and where the servant, which came first, "the chicken or the egg". The Center will be strong only when it rests upon mutually dependent economic links, and when these links are firmly established. Those who hold on to departmental rule by orders and decrees have not passed through any other experience. They are aware of only one management scheme, and sincerely believe that the whole problem resides only in its drawbacks and not in its complete uselessness.

World experience proves that large production and economic organizations (companies, corporations) have been operating effectively for many years within the framework of whole sectors. They ensure complete self-sufficiency, consistent growth as a consequence of stable profits, high rates of scientific and technological progress, and what is the most important, the full satisfaction of the market with a broad spectrum of products. Lately, such complexes have been acquiring the following qualities:

– increased diversification of production, regarded as the main factor in stabilizing the financial position of companies and corporations;

– increased orientation towards a long-term perspective, for which purpose forecasting and planning are intensively used;

– development of small businesses within the framework of companies and corporations, direct limitation of the size of enterprises and their orientation towards innovative activities and specific demands of the market;

– centralization of financial resources; allocation to affiliates and enterprises of only those means which are necessary for effective day to day management;

– organization of management on a matrix principle, coordinating manufacturing and production among regional companies and organizations;

– imposition of a collective character on the management of major industrial and economic complexes and in decision-making.

These features and trends were revealed with utmost clarity in 1989 during analysis of the organization and working methods of such companies as Shell and British Petroleum (Great Britain) whose character of production and output makes them close analogues to our oil-refining and petrochemical industries. There is a practical need to draw organizational and economic lessons from the experience of Shell and British Petroleum which are of immediate importance for the solution of the problems facing the restructuring of sectoral management.

Firstly, it can be taken for granted that many years of experience have proved the high effectiveness of such an organization in the operation of a large production and economic complex (company, corporation) under which: (a) practically all production, engineering, design, research, information and service functions are performed on a strictly cost-accounting basis; (b) all types of subdivisions and the complex as a whole operate on conditions of self-financing, self-sufficiency, clear-cut distribution of the responsibilities and rights in utilizing the resources; (c) all relations with state bodies – from local to superior authorities – are based on the taxation system and established regulations.

Secondly, it is of paramount significance that the overwhelming majority of the top level fundamental services, which in our conditions make up an apparatus of sectoral ministries, function in large corporations as "service companies". They operate on the basis of full cost-accounting, self-financing and self-sufficiency on contracts with production companies. This is the cost-accounting basis of the sectoral system as a whole which covers all its extractive, processing, servicing and intermediary enterprises and organizations. As a result, there are no "basic" and "superstructural" levels and a single, interdependent and inter-supplementing chain of cost-accounting companies operates to make up the whole system of their economic interests. Such a comprehensive "flow of cost-accounting" is accompanied by continuous formulation, desegregation and correction of corporate targets, depending on changing conditions of consumer demand, market relations, technological progress, priorities in different spheres of activities. The same basis is used to ensure flexible interaction among large, medium and small enterprises.

Thirdly, large sectoral and intersectoral concerns (including Shell and British Petroleum) have an efficient system of regional management and coordination which supplements (also on cost-accounting principles) the organizational structures of production and servicing. They are something like agents, local representatives or associations acting in the interests of the concern as a whole and financed by companies based in the given region. They are not a superior management body in the generally accepted meaning, but are territorial coordinators on matters of common interest to all companies in the given territory, such as taxes, prices, licences, patents, fines, relations with the government.

It is a fairly effective form of harmonization, of combining sectoral and territorial approaches to management, based on specially developed legal and economic mechanisms.

Fourthly, the cost-accounting foundations of the activity of the whole company predetermine the distribution of authority and responsibility at all levels (and not the other way round). In practice this means the setting of strict limits to the money each level is allowed to spend. Spending limits depend on the management level. In British Petroleum, for instance, there are 13 levels of responsibility measured by the sum (from 0.1 million to 410 million pounds sterling) which can be spent at each level. Hence, a clear-cut system of decision-making is assessed through profits and losses. And this system reflects the main interest of shareholders who are the starting point in the process of formation and accountability of structural sub-divisions and official positions, i.e. the board of directors, chairman, his staff, committees, etc.

Fifthly, radical changes in the organization of research and development have been under way in recent years in large productive and economic complexes, companies and corporations. Unfortunately, it has to be admitted that our practice of inclusion of research establishments into industrial and other associations, and all-round transference of research and development institutes to a regime of cost-accounting and self-financing reflects outdated ideas about methods of combining science with production. "Technological breakthroughs", costly technical designs, new science-intensive products and materials call for a concentration of the total intellectual potential, scientific, technical and material resources in key directions to win world markets. Efficient organization makes research work in present conditions highly effective. It has been estimated that every pound sterling invested in science brings 1.5 to 3.0 pounds worth of profit. Such is the case, for, example, with the Shell Corporation.

All this predetermines concentration rather than dispersal of research at the level of large productive and economic complexes. But now the main point is not a centralized hierarchy of research establishments, but a programmed and purposeful planning, financing and management of the total network of research laboratories and units to ensure the solution of key problems in scientific and technological progress.

The scope of this work is illustrated by the figures of expenditures on research and development in 1988 in the leading petroleum corporations of the world (in millions of dollars): Shell 721, Exxon 524, British Petroleum 381, Amoco 251, Chevron 249, Mobil 231, Texaco 170. Shell Corporation, for example, puts 85 percent of research projects into practice.

Joint goal-oriented research programmes with the concentration of all available intellectual, material and financial resources, cover not only central laboratories and research establishments, but also the local network of research laboratories affiliated with many companies and enterprises. The programmes are directed through setting of the targets, purposeful financing, regular control and coordination. But current local research work is carried out within the framework of production subdivisions at the cost of the respective production unit. In the light of such an agreement which takes into account the growing role and characteristic features of research in present conditions, it will be necessary to revise very seriously and carefully the views on the "cost-accounting of sectoral science" that have become widespread lately.

Sixthly, major productive and economic complexes – companies and corporations – cannot develop in the brutal conditions of competition on the world market unless they define long-term targets of their activities. Strategic planning and a "sliding plan" are the new instruments adopted by large scale economic systems. The former deals with global alternative scenarios of the corporation's development: exposure of possible problems, analysis of competitors and correlation of current plans with the strategy ("landmarks of development", "corner-stones"). They serve as a basis for developing the strategies of branches and companies, for estimating their contribution to the attainment of corporate targets. In the latter case, the required definitions are introduced annually into the plan for the next year with the addition of indices for the following five years. This is exactly what "the sliding plan" means – one year plus five, and the same goes on every year with a moving five-year period. The main advantage of such an approach is the practicability of planning with due regard for changes both inside and outside the corporation. The experience of the first Soviet concerns – TECHNOKHIM, ENERGOMASH, and QUANTEMP, forerunners of the new form of production management free from departmental control – convincingly proves that such economic organizations correspond in full to the conditions and requirements of the present stage of socialist economic development. The following characteristics make the company form of management highly efficient and promising: economic independence of enterprises; their voluntary affiliation with the company; democratic decision-making; accountability of the company to the labor collective which founded it; opportunities to manoeuvre resources, use them rationally and concentrate them in key directions; the cult of innovation and enterprise, etc.

It would seem that this path should make all the links of the economic system favor the functioning of companies, the more so as "the ice is broken", i.e. when dozens of companies are in the making. It must be clearly understood that the new form of economic management is incompatible with outdated schemes of centralized planning, the monopoly system of resource funding, and the sectoral principles of investment distribution.

Companies became "islands in the ocean" of old orders, business relationships, procedures and instructions. Ministries which let go of their enterprises suspended fulfilment of the long-term commitments that arose from governmental decisions taken earlier. Agencies providing materials and expertise stubbornly resist any change in the operation of their services. Planning and financial bodies refuse to take account of the specific conditions in which extra-ministerial structures are functioning. Various departments impose detailed and unnecessary book-keeping requirements upon companies, claiming the right to make decisions and place orders in the name of enterprises, disregarding the associate character and the essentially new type of industrial amalgamation.

The first companies were more successful in developing new relationships with state bodies, departments and other "external" organizations. It is the sphere of these relations that now constitutes the main barrier to the well-adjusted operation of economic organizations of the new type. As in other instances, the interrelation and interdependency of all elements of the economic mechanism move to the foreground. Domestic and world experience shows that the problem of effective functioning of large organizations should not overshadow the vast possibilities associated with the growing role and all possible developments of "minor forms" – small enterprises and small-scale production.

In 1987, the Russian translation was published of the book by T. Peters and R. Waterman, well-known American scholars of production management. Its title was "On the Way to Perfection" – this is how the authors expressed their intention to demonstrate the up-to-date level of organization and methods of management of American corporations. Scarcely had a few years passed, Peters published a new book, "Prosperity in Conditions of Chaos: Guidance for Making a Revolution in the Sphere of Current Management". One book essentially contradicts the other: from "superiority" to "stable chaos", from already achieved "perfection" to the need for a "revolution in management".

What is the matter? What is the reason for such a change of criteria?

Existence of a great many large corporations and companies is associated with the concentration of production and management, and with diversification. According to the author, these super-powerful structures, having taken monopolistic positions on the markets, directed their main efforts exclusively at ensuring their financial stability. And the problems of raising the competitive capacity and building up real market superiority were pushed to the background. The book refers to the practice of General Electric Corporation which, according to the author, lately has been seeking to raise the value of the company's shares and to speculate on the stock market, rather than to improve the quality and technical characteristics of its products. Since 1981, the corporation has acquired 325 enterprises worth 12 billion dollars and has sold 225 enterprises for a total of 8 billion dollars.

To counterbalance the"megalomania" which prevents constant change from having an effect on production proper, Peters calls for the instability of conditions and targets to be encouraged, and in this connection he points to new interest in the USA in small businesses. Small businesses have the potential to restore innovation and competition, to reinstate the priority of consumer demand. To keep an old customer is five times cheaper than to acquire a new one. The author believes that the future belongs to those companies which maintain permanent ties with their customers, follow their demands and are oriented to their unconditional satisfaction. To use his terminology, "stable chaos" is precisely a great multitude of simple organizational forms continuously renewed under the impact of changes in consumer demand.

This kind of organizational development of the basic level, when widespread in the national economy, must be enabled to work with the world trend towards changes in production and management structure, bearing also in mind the necessity to overcome monopolist trends by means of organizational measures, such as decentralization of production, establishment of competitive enterprises, and disbandment of ineffective associations of a monopoly type.

Narrow specialization, dynamism, flexibility of production programme, close relations with customers – all these allow small enterprises to make a sizeable contribution to the satisfaction of consumer demand. At present, there is no force in the country, except the ministries, that is capable of ensuring rational correlation of the large, medium and small enterprises, that are characteristic of the economies of other industrialized countries.

The vast possibilities hidden here are revealed by, for example, some data on the newly formed Ministry of Automobile and Agricultural Machinery. Large structural formations (state industrial associations, production, scientific and production associations) account for 53 percent, and independent medium and small enterprises for 47 percent of all industrial production. But in terms of proportion of workers, the share of medium and small enterprises is only 6.2 and 5.6 percent respectively.

Considering Soviet industry as a whole the proportion of enterprises with workforce below 500 is 70.5 percent, but their capitalization and gross production are very low: they account for only 10.3 percent of the capital funds, 14.7 percent of output, and 5.3 percent of the power consumed by industry.

In this connection, it is of interest that, according to the latest data, small firms of the USA employ nearly half of the workforce; over 37 percent of the gross national product comes from small business; most new ideas and innovative products are developed by new and small enterprises. It is also known, that approximately two out of every three new jobs are created by small business. Small enterprises account for 99 percent of the total number of American firms. Government programmes for the support and development of small business, as well as numerous publications on this problem invariably describe them in such terms as "the largest resource of America", "the driving force of prosperity", "the backbone of the US economy".

Small state, cooperative and lease-based enterprises are becoming prominent forces for improving the management structure of the socialist economy. Much is still to be done, with particular need for suitable legal, organizational and economic measures.

Reform of the organization of economic management – establishment of enterprises and associations, consolidation and development of modern management techniques are only a few aspects of the problem. Economic reform has brought to the foreground the problem of what function the sectoral ministries have, and how they should perform it.

It is not a question of whether or not there is a need for state organs of economic management in the socialist economic system. They are vitally important and necessary. The question actually is: what must be the tasks of these organs during the complex and difficult transitional period of restructuring the overall economic mechanism with the radical changes in the organization and methods of management that are necessary. It is not supervision of the current operation of enterprises or their replacement, but their management in accordance with their targets and investment, scientific, technical, and social programmes. Particularly they must ensure the needs of the national economy are met. These are the main differences between the functioning of the new type of ministries and that of the traditional ministries which we have for the present.

Intensification of economic reform, introduction of the socialist market, extensive development of direct ties among enterprises and accumulation of experience in regional cost-accounting will pave the way for further consolidation of state organs of economic management. Strategic and controlling functions will be strengthened, with the abolition of direct rule by orders. This course must be pursued consistently and unswervingly in order to overcome the cumbersome, complicated structure of state administration of the national economy, with delegation of many functions to lower levels, and an increase of dynamism and flexibility in the work of the top ranks of the executive.

This is why it is of paramount importance to redirect sectoral ministries from their function of management enterprises to the organization of the restructuring process, and as a top priority to create all necessary organizational and economic mechanisms for full scale cost-accounting and self-financing of enterprises. What is the practical meaning of this?

Firstly, it requires consistent, programmed work to reshape the basic unit (state and lease enterprises, cooperatives). It is important immediately to start setting up large production units (companies, corporations, industrial associations, inter-sectoral state associations, consortia) which will be able to ensure – on a cost-accounting basis with direct personal interest of all participants – the concentration of management, scientific, technical and other resources for the establishment of production. Having economic independence, these complexes will assume responsibility for the solution of a wide variety of problems previously in the competence of ministries. They lay the foundation for the future transition to essentially a non-ministerial system of production management.

For the same purpose, establishment of the system of cost-accounting, in parallel with that of large-scale organizations of a corporative type, will be spread over the sectors themselves characterized by shared technological processes and close internal cooperation in the manufacture and marketing of the final product.

World experience related to functioning of large production units as autonomous cost-accounting and self-financing systems suggests the regime which governs the functioning of state corporations should be introduced, implying:

– transition to cost-accounting and self-financing of a whole economic complex which many include economically independent, and differently managed, organizations which in the aggregate make up a sector or a sub-sector of industry;

– development of the complex's management system on democratic principles with such institutions as shareholders meetings, a board of directors, management, etc., with distribution of authority among them.

It has been established that such sectoral cost-accounting complexes (state corporations) could successfully operate in the complete cycle of extraction, transportation and distribution of natural gas, in the gold-mining and diamond industry, in the oil refining industry, in the production and marketing of fertilizers, etc.

The principle of voluntary participation on equal terms of enterprises and associations must be strictly adhered to. It is important to draw lessons from the past which have shown substantial losses to accrue in the course of restructuring sectoral management organizations. For example, state industrial associations whose formation was really intended to revive the middle link in sectoral management are known to have been established by order, through artificial unification of groups of enterprises, with complete disregard of their interests. The results were deplorable.

Secondly, radically restructuring the work of ministries and shaping essentially new state sectoral and intersectoral management organs, necessitates consideration of their past function as controls of the command and administration system. The monopoly position of almighty departments which commanded the fundamental sectors of the national economy allowed them unchallenged control over the economies of Union republics. Taking advantage of their dominant position in the national economy, they frequently neglected the interests of republics and regions, which often had imposed upon them a ministerial decision which ignored their local conditions and their prospects of development, and local public opinion. This gave rise to discontent over the actions of central government, and to inter-ethnic tensions.

This must be radically changed. A change in the relations between ministries and economically independent enterprises involves a transition to republican cost-accounting, to decentralized formation of the budget system, and to substantial consolidation of republics' sovereignty. Further, the competence of All-Union and republican organs must be changed to conform to contemporary realities of the greater independence of provinces, taking due account of the interests and peculiarities of each republic of the Union.

The interests of economic sectors and of the republics and regions must be coordinated. This may involve many channels, including elaboration of plans, distribution of productive forces, definition of the profile of newly-built enterprises and those which are undergoing reconstruction. It is no less important to seek mutual benefit in the development of production and social infrastructure, utilization of labor and production resources, and the improvement of the environment.

Thirdly, the role of ministries in the realization of restructuring processes must be made very obvious in the consolidation of economic reform, the consistent democratization of management, full cost-accounting, self-financing and autonomy of enterprises. While delegating many functions of management down to enterprises, sectoral ministries must guarantee economic independence and rights. Supplying enterprises with initial planning data (including state orders and quotas) the ministries are expected to shape step-by-step a new system of economic relations based on utilization of economic levers, observance of mutual profitability and legal commitments, development of competition and direct ties between enterprises, horizontal integration of enterprises, etc. Ministries should encourage enterprises to withdraw from their jurisdiction and to switch over to leasing, or to convert into cooperative and stock-holding enterprises with membership of inter-sectoral state associations and companies.

Hence, departure from strict administrative subordination and jurisdiction of enterprises will be inevitable. To manage through service, to lead through assistance, to control through rules – this is what the ministries must have in mind in their relations with enterprises. The task is completely to overcome the concept of "jurisdiction" which defined the administration-dominated hierarchy of management. It is the labor collectives – the productive units, rather than administrations, that are to be the new holders of power. And enterprises which enter into relations with ministries and other bodies on the basis of equality and partnership must be subordinated only to labor collectives and to law.

The more complete the economic reform, the more widespread must confidentiality of contract details become, taking into account the mutual interests of the sides. At present, confidentiality of contract is being established between ministries and extra-departmental companies and associations. In future, it will be characteristic of relations among any enterprises, including those which are affiliated with respective sectoral systems.

Such an approach to the organization of management will allow the ministries, in pursuit of the targets of national economic and social development, to concentrate in practice on their proper functions, such as direction of investment, strategic research and development, encouragement of cooperation within and among the sectors, relocation of enterprises to suitable regions, assistance to the maturing of the socialist market. Solution of the problems at national economic, sectoral and regional levels is a main aspect in their activity.

To meet the new challenges, the state organs of economic management must be equipped with three main instruments:

(1) programmes of national significance;

(2) orders for carrying out the programmes (with control of all economic levers);

(3) control over effective functioning of the cost-accounting sphere.

Many functions such as productive and technical supplies and sales; technical, information and legal service; training and re-training of personnel have previously been performed by the central authority and these must be replaced by cost-accounting contracting organizations to service enterprises. On this path, the old system of merging state administration functions with economic management functions must be resolutely and uncompromisingly overcome.

Chapter VI

The economy of a region: planning, management, cost-accounting

B.Z. Milner

Professor of Economics, Acting Director of the Institute of Economics, USSR Academy of Sciences

A peculiar feature of the present stage in the development of the socialist economy is that the further socialization of all stages of production requires new forms of organization of economic activity covering each geographic unit and the whole national-economic complex. As major industrial-economic complexes and various kinds of horizontal integration of the main links emerge, establishing the territorial level of economic activity, and turning regions into relatively independent economic formations, is acquiring primary importance.

One must be aware that territorial cost-accounting, as introduced recently, is not absolute. One can speak of cost-accounting and self-financing in the proper and complete sense of the terms only in relation to a commodity producer – the main unit of the economy. But when we speak of a region, of a union republic, then we mean in practice that the quality of life of the population and the volume of resources created by its own labor must be in relation to each other.

The resolution of the XIXth All-Union CPSU Conference "On International Relations" says: "We should organize everything so the working people understand clearly what the volume of production in a republic or a region is, what contribution into the economy of the country it makes and how much it earns. The idea of a transition of the republics and regions to the principles of cost-accounting with an accurate estimation of their contribution into the implementation of the All-Union programmes is worth our attention".

The switch-over of the whole system of managing regions and republics according to the principles of cost-accounting should ensure their independence in using their resources, in working out and successfully implementing socio-economic programmes, and should permit independent action and initiative in deriving profits, thereby ensuring a high standard of living for the population. Every region, and every republic should be interested in improving its economy to assure its prosperity, thus multiplying the common wealth and might of the Soviet state.

1. Basics of Regional Economics

The united economic complex of the country is based on the nationwide processes of socialization of production, specialization and cooperation, and of the All-Union division of the collective labor of society. Many factors influence these processes – specific features of the branches of industry, the existing distribution of productive forces, capabilities of technological systems, the nature and value of the commodities produced, natural, historical, demographic and other features unique to each region. Under such conditions the united national-economic complex cannot be just an arithmetic sum of the economies of the regions. The whole of the industrial-technical potential and the material-resource basis determine the economic independence, stable and dynamic development of both the USSR as a whole and each of the republics in particular.

Turning regions into relatively independent economic formations is possible provided they are components of the united national-economic complex and at the same time they themselves represent an integral regional national-economic complex. It is conditioned by logical processes in the development of the public organization of production, including the territorial division of labor, specialization and cooperation of production.

The territorial division of labor is directly connected with the division of labor in general, and it determines the specialization of regions and the emergence of inter-regional economic links. The specialization of regions means giving up self sufficiency and isolation. It is aimed at developing mass production of certain goods or groups of goods for exchange with other regions which are themselves specialize in different ways. Thus the specialization of the regions ensures their organic involvement into the united integrated national-economic complex, and determines the volume of the region's contribution to the economy of the country and to its own development.

The territorial division of labor and specialization of the regions is only one factor determining the place of the region in the economy of the country. The other factor is territorial cooperation of production which makes the regional national-economic complex self contained within its specialization. A region should be considered not merely as a location for factories, but also as a large-scale combination of various natural material and human resources, connected with industrial projects by a complicated system of stable mutual links.

The existence of a sum total of various stable interconnections, comprising the productive and territorial unity of interdependent projects, forms a region.

Definition of regional boundaries must be in terms of economic structures whose components are the union and autonomous republics, the territories and regions, the cities and administrative districts. The effectiveness of the national-economic complex of the region is determined by factors among which is the optimum combination of enterprises, the improvement of the structure of production, the usage of local natural and labor resources, the protection of the environment, the training of the management and workforce.

The effectiveness of the regional national-economic complex is directly connected with integrated development for communal use of the regional infrastructure of production.

The development of social infrastructure, planned use of local opportunities for increasing the efficiency and volume of production of foodstuffs and consumer goods are of special importance.

At the present stage in economic development direct productive links between enterprises in the same territory, and the organization of joint activities to achieve solution of economic and social problems are seen as indicating the successful functioning of regional complexes. Under the conditions of an emerging market economy the formation of regional markets of commodities and services, the development of a corresponding infrastructure are of great importance for the growth of efficiency.

One can single out the following types of regional complexes: agro-industrial complexes within administrative regions, industrial complexes within the regions and autonomous formations, complexes based on the union republics, and complexes of major economic districts, embracing a number of regions and even republics.

The development of economically developing or underdeveloped territories brought about the emergence of territorial industrial complexes (TIC). TICs are established in order to make it possible for the enterprises of several industrial branches to find joint solutions of economic problems on the basis of developing natural resources of poorly, or patchily developed districts. We mean such TICs as Kansko-Atchinsk, Pavlodar-Ekibastuz fuel and energy complexes, the TICs of the Kursk magnetic anomaly, the West-Siberian oil and gas complex and others which are being formed with centralized capital investments.

At present the planning of comprehensive economic and social development of the regions is undergoing radical changes.

First, these changes require the realizing of huge and so far poorly exploited resources, raising the efficiency of public production, and the accelerated solution of economic and social problems. Republic and regional plans must give up their former role as being consolidated and summarizing documents and become effective instruments of defining and regulating the processes of optimizing territorial division of labor, and of strengthening territorial cooperation of public production.

Second, territorial plans change their nature in connection with the switch-over to the use of primarily economic methods of management of economically independent enterprises. Territorial plans should set the tasks of developing a regional national-economic complex taking into account the interests of the country in general and should determine the ways of carrying out those tasks on the basis of using economic regulators. Such regulators are state and regional orders, local rates of pay for labor and prices for natural commodities, taxes and tax privileges, subsidies, grants and fines.

State orders are to be placed on territorial basis for the union republics (to supply vital foodstuffs and agricultural produce to All-Union stocks) and for ministries and departments of the USSR on other types of produce determining the specialized division of labor of the republics and districts. Together with state orders, and in order to ensure their fulfilment the republics are given limits for centralized capital investments and material and technical reserves. As the economic independence of the main enterprises and amalgamations strengthens and the socialist market is expanding, the number and volume of state orders placed on a territorial basis will be reduced.

The use of scientifically based territorial planning for effective and balanced development of the republics and regions will enhance the importance at the All-Inion, republican and regional levels of regional forecasts and pre-planning research on the development and distribution of productive forces.

Regional programmes of socio-economic and scientific and technological development should be used as a major instrument of planning under the new economic mechanism. Fair play in inter-regional economic relations and a guaranteed minimum standard of services in all regions for all citizens of the USSR should be ensured through the use of planning levers.

2. Restructuring the Management of the Economy and the Social Development of the Regions

A new approach to territorial management is being implemented in fundamental restructuring of national economic management. Principally, the economic independence of the union and autonomous republics, and of other national formations is being strengthened, and the development of the regions is being promoted, with favorable conditions for business management on the basis of complete cost-accounting and self-financing being created in all the regions.

The management of the economy and the social development of the regions is being restructured so as to ensure a new standard of living for the population, with a modern level of productive and intellectual activities, and the highest standards of functioning of the territories, through effective management promoting economic independence.

This implies solution of the following problems:
– in the field of social development – the creation of conditions for the most complete satisfaction of material and spiritual needs of the population, guaranteed social justice in the distribution of the disposable income, improvement of labor conditions, living and recreational conditions;

– in the field of economic development – effective use of the scientific, tech-nological, productive and intellectual potential of the region, the introduction of a modern system of businesses, integration of the economic interests of the regions, the enterprises and the population, further specialization of the regions regarding labor and the formation of a united socialist market;

– in the field of management – a clear distinction of the functions of the All-Union, republican and other territorial organs as far as the economy of the region is concerned, the creation of an efficient structure for territorial management;

– in the field of employment and demographic development – the creation of permanent yet flexible employment of the population, the stimulation of its growth, and the regulation of migration processes.

A number of requirements must be taken into consideration when switching over to these principles of territorial economy.

First, the new economic and legal conditions must strengthen the legislative authority of the elective bodies, namely the Soviets of People's Deputies at different levels, while preserving the unity of the region as a single socio-economic unit.

Second, the new economic independence of territories must not infringe on the cost-accounting rights of businesses, but it should ensure their com-plete implementation. The relations of the regional organs with enterprises and amalgamations must comply strictly with the laws of the USSR as they relate to property, land and land utilization, leasing and leasing relations, the uniform tax system, socialist enterprise, and general principles of managing local government, the economy and society in the union republics.

Third, restructuring the management of regional economies presupposes development of inter-territorial links, expansion of the economic relations between different regions, with maximum usage of the advantages of the specialization of individual regions and the All-Union division of labor.

The fundamental basis for these developments is definition of the forms of ownership of property as they exist in different levels of the socialist national economy – All-Union, republican, and local. The plurality of the kinds of property owned must also be recognized. By that we mean that in order clearly to determine rights and responsibilities of ownership and custodianship of property, the affiliation of industrial, transport, agricultural, trade, communal and other enterprises to types of socialist property should be established. Ways of regulating the relations of property are to be agreed and established on the All-Union scale.

The law states that property might be in the name of a district, a city, a region, a territory, a republic provided the property itself does not cease to be socialist. In any case the objects owned are to be defined in terms of stages in socialist production. Singling out this or that item of property must be accompanied by a statement of its extent: if any objects of property are made over to the republic, one should be at the same time aware of what objects still remain All-Union property and which are to be given to the local Soviets. Endowing republican and local Soviets with property from the socialist stock lays an economic foundation for their political power.

With that in mind, the united fund of public property is to be distributed first of all among the Union of SSR, the union and autonomous republics, and local Soviets of People's Deputies on a democratic basis. State property acquires a really national nature only as a result of distributing the means of production between the Union of the SSR and the union republics. The delimitation of property takes place at different levels: the Union of SSR, the union and autonomous republics, local Soviets of People's Deputies.

The democratization of the management of the national economy requires the updating of all tiers of management. It has become necessary to grant and legally to consolidate to union republics and regions the new rights of the being in charge of property and natural resources. This is a decisive prerequisite for restructuring the management of the economy and society on the basis of self-government and self-financing.

The elected representatives (or the instruments of government under their control) should perform the duties of the guardians of socialist property – to guarantee its maintenance and growth by appropriate management. The use of such new forms of management as leasing, cooperative ownership and joint-stock companies can turn what was purely state property into property of the people while preserving the role of the state as one of the owners of property and as the economic center.

Those responsible for state property are:
– in the case of productive enterprises and economic organizations at the All-Union and republican level – the corresponding union and republican executives;
– in the case of enterprises producing consumer goods, of industries servicing territories, and bodies controlling other material and natural resources (land, water, public buildings and works, etc.) of territorial usage – the corresponding Soviets of People's Deputies;
– in the case of local enterprises and economic organizations, and other material and natural resources of local importance and usage – district Soviets of People's Deputies.

Soviets of People's Deputies or their executives are empowered to enter into economic relations with labor collectives of state enterprises and organizations, cooperatives, joint-stock companies, and individual producers which may then, under the law, own or have the use of socialist property under an agreement, lease or contract.

Diversification of forms of ownership will ensure the most effective use of property under present-day conditions and this has become the foundation of a considerable increase in economic independence of the Union republics. This new philosophy is a consequence of the democratization of all aspects of public life, by the move towards the comprehensive use of resources and the development of production, by rectifying the distortion arising from the centralization of the management of economy. The underestimation which has existed for many years of regional and local opportunities for development of the national-economic complex of the country must be overcome in this process.

Lenin's understanding of the paths for the development of the regions in a socialist state is again becoming topical. "... In the same way as democratic centralism does not at all exclude autonomy and federation, it does not at all exclude but, vice versa, presupposes a complete freedom of various localities and even different communities of the state in working out diverse forms of state, public and economic life. Nothing can be a greater mistake than mixing up democratic centralism with bureaucracy and conventionalization."[1]

Economic independence cannot be granted to the republics without reference to all the other elements of radical reform of the economic mechanism. Three fundamental factors must be considered: firstly, the increase in the economic independence of the main businesses (the enterprises and amalgamations of various types) and the completeness of their switch-over to cost-accounting and self-financing; secondly, the balanced development of the national-economic complex of the country and the degree of economic independence of the region in so far as they are connected; thirdly, the necessity to make the economic independence, rights, and responsibilities of a republic conformable with management on the country-wide scale. A pragmatic approach will be necessary in taking these factors into account according to local conditions in this or that republic.

The completion of economic reform which has as its basis complete cost-accounting and self-financing of the main businesses clearly requires a proliferation of horizontal ties, with organizational integration of management among businesses and within the territory. The reorganization of production through creation of enterprises such as: those independent of government departments, socialist companies, inter-republican and inter-business amalgamations, joint ventures with foreign firms, consortia, and other progressive forms of business organization is to be the direction taken by the re-organization of the economic activity in the republics.

---

[1] V.I. Lenin, Collected Works, Vol. 36, pp. 151–152 (in Russian).

The break with the branch (departmental) principle which worked by administrative subordination of businesses, makes possible the economic independence of the union and the autonomous republics, thereby creating normal and legal relations between local authorities and businesses.

Regional economic management must not be established simply by a switch-over of administrative powers from one level to another – from the union level to the republican or local. In practice it means the substitution of the departmental administrative system by an analogous regional system.

Interrelations between businesses and local management must not be on the basis of administrative subordination but rather on the basis of the legislation determining the participation of the enterprises in forming the financial foundation of the republics and regions, and in using of natural and labor resources and the social infrastructure. The union, autonomous republics and other national-territorial formations must be granted maximum independence in the organization of social and economic life in order to develop the cultures of our peoples. It is important to ensure a social orientation of productive and economic activities, and to raise to prominence the economic significance of environmental protection and the wise use of resources.

Taking into consideration these and other conditions we are implementing the organic combination of the interests of labor collectives and the interests of the regions, with the inter-connection and inter-dependence of the application of cost-accounting principles in the activities of enterprises and on the territorial scale.

3. The Fundamentals and Methods of Regional Cost-Accounting

Today regional cost-accounting is becoming one of the important economic methods available for management of the national economy.

In the process of reforming the political system, moving to the rule-of-law state and restructuring the organization of management, the rights of local and republican Soviets are expanding, with their material basis being consolidated. This enables them to make a real contribution to development of every territory, and of the country as a whole. The Soviets won't be able to act as the political basis of our state unless they enjoy economic power. On the other hand, considerable powers which previously belonged to central government are being devolved to them. All necessary measures are being taken legally to consolidate the clearly distinguished functions and rights of the union, republican and local managements.

A principle for distinguishing the spheres of the activities and functions proper to the union, from those proper to the republics must arise from the concept that the independence of the union republics and their responsibility for meeting various needs of the population should be considerably broadened. These needs include housing, consumer goods, service industries, social and welfare provision. That means that the material and technical basis of the consumer sector of the economy must be managed directly by the republics, that is, we must transfer businesses of agro-industrial type, light industry, elements of the construction industry, the services and the production of consumer goods. Projects relating to the means of production of regional importance and inter-branch enterprises must be managed by republican and local government. Administration of housing facilities and public utilities, local transport, health care, education and other establishments servicing the population are to be subordinated to the regions.

Only the functions which stem from the organization of the All-Union division of labor should be maintained at the national level. The prime influence of central government in solving these problems should be the placing of state orders and the planning of major investment, scientific and technological programmes, which are important for more than one union republic.

The following examples illustrate the prospects opened up for raising efficiency of the national economy.

The excessive centralization of the management of the economy, and the dominating role of the departments have led to the current situation that republican industry accounts for less than 5 percent of total production. The republics were in charge of only one fifth of the state centralized capital investments.

According to preliminary estimates republican industry will produce approximately 80 percent of the total volume of consumer goods. In general the share of the republican industry will increase 8–10 fold while in the Baltic republics, Central Asia and Moldavia it will comprise 60–75 percent of the total industrial production. The share of the republics in state centralized capital investments will increase 1.5–2 fold.

At the All-Union level, management in the union republics uses, primarily, economic methods. Reducing the number of administrative layers, simplifying the management of enlarged business, changing to ministries of a new style must all be functionally correlated with restructuring the management of the economy of the republics. The switch-over to new forms of management must not lead to excessive independence of the territorial economy. Administrative procedures must not be discarded only to be introduced at the republican level. These structures must not interfere with, but rather promote the organizational integration of major businesses irrespective of their departmental or territorial subordination. They can be formed on a branch, inter-branch and inter-territorial basis reflecting natural laws of the development of productive forces.

The precise pattern of cost-accounting, economic methods and forms of management may differ. Though the basic principles are the same, they may be put into effect differently, according to the republic. The degree of socialization of production and the structure of the productive forces will influence the pattern. It is also necessary to take into account economic efficiency and social justice. Regional cost-accounting and the economic independence of the regions will be a vital component of the new, national economic mechanism. This mechanism embraces economic relations between: a) the local Soviets and all the business in that territory; b) the Soviets and local, republican and union managements; c) the Soviets of various regions as they may collaborate to solve general problems of the socio-economic development of the territories.

There is a basic distinction between cost-accounting as it applies to a business and as it applies to a territory. The gist of it is that cost-accounting is a feature of commodity production and in the proper sense of the word one can speak of cost-accounting only in relation to the commodity producer, an enterprise. Cost-accounting is a system of production relations which arises from the economic independence of a commodity producer. Earning money through the production and marketing of goods or services, refunding production costs (self-repayment), developing production (self-financing) all confer independence in taking management decisions connected with the use of the money earned (self-government).

As for the region and, even more so, for the Soviet of People's Deputies as management executives it would be absolutely wrong to consider that they have to "earn" their income in the same way as commodity producers. Of course, a region cannot be considered to be a commodity producer. A region is a complicated complex of independent producers and this is another reason why the Soviet of People's Deputies cannot be treated as a commodity producer. The activities of any organ of state power and management, whether a local Soviet or an All-Union ministry, cannot be considered as commercial, as requiring payment for the management services they provide. In this respect it is not correct to speak of pure territorial cost-accounting or cost-accounting of a ministry.

One cannot, of course, deny that territorial governments might carry out commercial activities, perhaps earning money, by leasing their property or investing in joint-stock companies, etc. But commercial activities do not define the economic rights of the Soviets of People's Deputies as organs of state power.

Other production relations differ qualitatively from the relations between commodity producers and these define the incomes of the Soviets of People's Deputies. Deriving profits from the resources belonging to society is a realization of socialist property, the sovereign owner of which is the government of the people as a whole. That is the key distinction between territorial cost-accounting and enterprise cost-accounting.

Cost-accounting relations between the territorial economy and enterprises must be based on the following fundamental principles: reimbursement, with the relations between parties being based on an equivalent exchange; mutual benefit of transactions between the region and enterprises; responsibility of the transactors for any breach of contract. The change to that system must not precede, but follow or take place simultaneously with the introduction of real enterprise cost-accounting.

The use of new principles in forming republican and local budgets is of basic importance. In earlier periods the budget was made up of expenses envisaged, to cover which the necessary means including subsidies were produced. After the switch-over to self-financing the budgeting of the republics, cities, and districts will be carried out on the basis of the amount of the money earned by this or that region. At the initial stage subsidies can also be used.

The income of the union republics will arise from: the profits of the subordinated enterprises excluding that part transferred to the union budget; the payments for land, water, labor resources; state taxes paid by the population; all local taxes and dues; export profits. A union republic plans and balances its budget independently using such economic levers as prices and credits.

Not only the business which are subordinated to the republic will contribute to the budget, but also other enterprises located in the territory and subordinated to ministries or All-Union companies, associations, amalgamations. An enterprise owned by a union, located in a republic and using its labor and natural resources must participate in financing its social and productive infrastructure. Contributions of the enterprises into republican budgets (tax from profits), payments for natural and labor resources, will make up a considerable part of the income of the republics. The residents of a republic irrespective of the nature and remuneration of their work must enjoy all the social benefits created on its territory.

All kinds of activities should be assured of financial support from the budget of their corresponding Soviet, and of reimbursement of the costs incurred by: a) meeting the needs of the population (material, spiritual, social); b) ensuring regional conditions for an effective operation of all the enterprises in the region (production, supplies, infrastructure, etc.); c) preserving historical and cultural monuments; d) maintaining the quality of the environment.

Effective inter-regional relations are important for the applying of cost-accounting principles and creating economic independence of the republics. In the framework of the development of the All-Union market the terms of economic exchange between enterprises and organizations as well as between the republics must be determined on a contractual basis. Because of uneven development and the necessity of a structural reconstruction of the economy we have to face the fact that some regions will lag behind, and advanced republics should render voluntary assistance. All republics participate in creating an All-Union fund for supporting underdeveloped regions as well as those which are victims of natural disasters, or which are developing new territories.

According to official statistics, only two out of the fifteen union republics have a favorable balance of the trade in inter-republican commodity exchange. The use of one or two indexes in assessing such complicated processes does not reflect the real system of economic interrelations between the republics and may lead to incorrect conclusions. There are various proposals as to the methods of calculations, but no uniform approach has been worked out so far. The assessment of imports and exports of goods must, of course, take into consideration various price proportions and changes, correlations between the exchange of domestic and foreign-made goods, the volume of export–import operations and other factors.

One of the characteristic features of economic independence of a republic could be, for instance, the correlation between the produced and the disbursed national income. Using such an index makes obvious the degree of the utilization of domestic resources for socio-economic development. But one cannot disregard the justified redistribution of the national income between the union republics for the purposes of stable and balanced development. This makes calculations of a correlation between the produced and the disbursed national income not fully correct. Besides, we should also take into special account the redistribution of funds in connection with local features of the investment policy and economy in the republics and in the interests of the country as a whole, as well as the need to render international assistance (for instance, in connection with the earthquake in Armenia, the catastrophe at the Chernobyl atomic power plant in the Ukraine, etc.).

Under the conditions of the socialist economic system an important task of the union organs is to help the republics to combine their efforts and resources in bilateral or multilateral projects. When necessary, inter-republican commissions on economic, scientific and technological and cultural cooperation should be established. It should also be taken into account at the same time that the establishment and organization of mutually beneficial inter-regional links is directly dependent on the solution of such key problems of the economic reform as the development of commodity and money relations, wholesale trade, and the formation of a united union socialist market. The strengthening economic independence of the republics, the growth of their national-economic potential, and the growth of their well-being are inseparable from the process of deepening specialization, economic integration, and build-up of their common scientific and technological potential.

The Supreme Soviet of the USSR has granted economic independence on the principle of cost-accounting to Lithuania, Latvia and Estonia. These principles mean a considerable broadening of the republics' authority in their finance and crediting activities, in taxation, price setting, and in using their incomes. Similar principles will become the foundation for the development of economic independence of other union republics as well. The very notion of economic and political sovereignty of a union republic is becoming real and full-blooded. Going beyond the framework of the economy proper, the switch-over to these principles opens up vistas for considerable progress in all aspects of life, for further development of democracy, and for increasing the well-being of the population of the republics.

Chapter VII

Socialist ownership: from uniformity to multiformity

L.V. Nikiforov

Institute of Economics, USSR Academy of Sciences

1. Moving from Nationalization to a Civic Society

1.1. After effects of rejecting alternative ways of development

For many decades direct or indirect rejection of alternative forms of social and economic development in general and of socialism in particular has been a tenet that ranked prominently in the USSR's prevailing ideology. Featuring an outright ideological intolerance the tenet was a basic alienating factor, prompting confrontation and overt enmity even among essentially similar political movements, let alone among fundamentally differing theories and practices of social development.

The rejection of reasonable alternative paths to social and economic progress was encouraged by simplified and caricaturized views of social and economic development. The tremendous diversity of patterns of evolution and history of social systems and structures was reduced to five basic sets of social and historical phenomena or systems that were expected, or predestined by some divine power, to culminate in Communism, as proclaimed by the slogan "The Victory of Worldwide Communism Is Inevitable", which was a sort of revived world revolution concept. The possible emergence of other social systems resting on criteria that differed from those of social class was simply ignored and optional or alternative patterns of evolution were ruled out. The social and economic relations inherent in each system were presented as rigid schemes. Such schemes hinged on an impending change-over from capitalism to socialism (communism) and a predominantly "converse" relationship between capitalist and socialist systems as regards social and economic relations.

At the same time, it was recognized that at the initial stages of socialism the "birth-marks" of capitalism and the survivals of bourgeois law would stay with us as phenomena affecting our new society from the outside, and alien to its intrinsic form. As a matter of fact, most of the real contradictions inherent in social development were viewed as foreign to socialism. Such an approach would, willy-nilly, firstly, bring forth a socialist scheme divorced from reality and, therefore, incapable of translation into reality and, secondly, reject the likely emergence of new social and economic relations and systems themselves.

The descriptions of evolving social systems would take no account of capitalism and socialism as industrial societies sharing the same material and technological underpinning and, as a result, would ignore the significance of the developmental trends and social structures that are predetermined by their productive forces. Therefore no consideration would be given to the thrust and substance of social changes that affect the quintessential dimensions of capitalism as it advances toward new stages of industrial and post-industrial development. No attention was paid to the fact that the capitalist system underwent transformations in an effort to adjust itself to completely new productive forces as well as to the choice of fundamentally different and new alternative strategies for social and economic progress. In the final analysis, what was at first unnoticed against the revolutionary backdrop and which emerged as an increasingly obvious feature of productive forces was the crucial fact that as society continued to advance, its basic social and economic options and alternatives increase in scope, for objective and subjective factors grow in number and interact in their own peculiar ways to affect society's progress. As a result, more intricate relationships emerge, and productive forces and their social and economic environments modify each other in the process of linking up. This gives rise to the evolution of fundamentally different social and economic structures and systems as well as to a multiplicity of versions (including alternative options) in each structure and system, which results in physical, social and intellectual progress. Historical experience demonstrates that any system and its versions that lend themselves well to self-regulation are quicker to take advantage of the vistas opened by progressive shifts in productive forces. This sets the stage for the continued growth of material and intellectual culture, for such a system is capable of responding to a greater number of developmental factors and proves to be more dynamic and sustainable.

But the societal development scheme that had been devised and approved as an ideology disregarded all those factors and provided, for purposes of producing and selecting appropriate versions, only a limited number of specific ways for translating the new social system's general trends into reality, most of those trends having been borrowed from Soviet experiences, or postulated in a speculative manner. Another aspect of version selection stems from the conclusion that some nations may avoid capitalism because they continue to stay in touch with the world socialist system. This appears to have been the only setback suffered by a system that otherwise is forging ahead relentlessly.

Some divine power, it appears, set down the one-dimensional objective that society is supposed to develop towards. Further, only "our" system is based on a true understanding and rationalization of this concept. In fact, when we ponder this, we are forced to conclude that our rationalization actually stems from the system wanting to protect itself against the impacts of the outside non-linear and multidimensional world of social relations, ideas and ideologies. The system had reasons to do that, for it was based on the power of its ruling elite and on total centralization and nationalization permeating all walks of life in our society and, as result, it did not have sufficient internal resources and incentives for development. In the absence of such resources the system attempted to seek cover behind a shield of ideology by claiming that its social design was the only correct one and that best suited to keep abreast of history's pace. The price we have had to pay for such a protective cover has proved exorbitant.

It is hard to appreciate the social, economic, political, cultural, scientific, moral and ethical damage that has been done by the long prevailing concepts of societal development as well as by the established system of societal relations that claims to serve as the only correct model of socialism. There is no denying that these concepts have provided fertile breeding grounds for division and sepa-ratism in socialist and communist movements, disrupted international economic, scientific, technical, cultural, political and other ties, and proved one of the reasons for missed opportunities in the field of production advances, social and intellectual progress, and numerous major gains of human civilization that could have been put to good use. In the final analysis, the practical implementation, and at times the overt imposition, of the "monomodel new society" version on a large number of countries set the stage for the gradual development of grave economic, social and political crises there and for the diminishing prestige of the socialist idea that for no good reason, was identified with its oft-deficient materialization. In methodological terms, viewing historical evolution in general and the evolution of socialism in particular with its monolinear objective-setting is but a departure from a truly materialistic interpretation of history.

1.2. Fundamental diversity of social and economic development

The world is undergoing sweeping changes now, with a revolution in science and technology in capitalist countries, and the evolution of capitalist relations that make it possible to tackle a number of social problems. Countries that have embarked upon socialist construction are facing hard times and the cardinal social, economic and political changes that are sweeping across many such countries have combined to confound the simplified theoretical conclusions picturing today's realities in black and white colors. Such a naive view regards our age as an age of gradual transition from capitalism to socialism according to what has been described as a dominant trend in world development. This naive view is clearly completely refuted, together with the utterly untenable concepts of monomodel socialism.

The simplified concepts of one, or to be more precise, a singular model of socialism stemmed primarily from some abstract scheme that watered down and distorted our vision of reality by incorporating general abstract objectives of social development and several equally general concepts describing the trends and ways of attaining those objectives. Secondarily, attempts were made to stretch diverse patterns of social development and their combinations to fit the procrustean bed of a "standard model" scheme which incidentally was never translated into reality and was used as kind of smoke screen to camouflage the realities that were in fact worlds distant from that scheme.

The conclusions regarding alternative versions of social and economic development in general and those of socialism in particular mirror the diversity of conditions that obtain in various countries, how those conditions change from time to time , and, lastly, the expansion in many dimensions under the impact of many factors of social and economic developments that are unlikely to fit in with uniform schemes. As a result of the appreciation of the multidimensional scope of social development attempts were made to identify various versions of evolving socialism and alternative development that would be appropriate for these developing social structures and economic patterns, ownership relations, economic machineries, etc. while some countries saw campaigns mounted again in favor of non-socialist development. This makes it imperative for us to examine the fundamental versions and alternatives of socialist models and to devise new concepts describing socialist criteria.

In this connection it should be pointed out that fundamental social development alternatives may not be reducible to just one reason or factor. Comparisons of the experiences gained in devising such specific development versions suggest that alternatives have wide conceptual bases that permit growth of a diversity of alternative versions which can develop at each stage of evolution and in each country. Therefore the basic alternatives ought to be classified to enable us to visualize the major directions that those alternatives might take. It appears that basic alternatives of social and economic development are engendered mostly by, though are not reducible to, the diverse trends of evolving production and its true socialization. At the same time, what is in question here is not only the specific features of such developments as shaped by differing evolutionary and social factors in various countries and economic sectors. Such specific features are likely to level off bit by bit. The point is that it is impossible to make uniform the modes of socialization. The ways whereby socialization really happens, the distinctive links that are established as a result, are made contingent upon various factors. These include how such factors relate to the productive forces that are fundamental to a social structure and, therefore, generate a need for a diversity of ways of owning and profiting from the economic structure. Such productive forces feature above all those based on the immediate exploitation of living nature's provision of the means and tools for work, as well as those that use living nature only as a general prerequisite and a working environment. It is impossible to escape the characteristics of those types of productive force. They shape the specific modus operandi, life-style and living standards of the social groups of people involved.

Productive forces can differ very greatly with their traditional and brand new elements in stark contrast to one another. A case in point is how information systems, the spread of knowledge and other phenomena can usher in fundamentally new ways of enhancing and effecting production and of establishing interaction between scientific, technological and economic factors to contribute to economic growth and to add to, or detract from, the efficiency of social structures.

The specific productive forces that influence social and economic relations also depend upon which resources are available, and other factors also influence distinctive patterns of production.

And lastly, alternative forms of social and economic development can involve changes to productive forces and a correlation between the direction of real-life production socialization and its mode. Diversity in productive forces and in physical assets to be appropriated as well as changes in their nature dictate the necessity of a diversity of ways of socializing production and its economic benefits in practice. This diversity ranges from one's individual property and business to public ownership, including all kinds of combination including territorial groupings. Today's trends of the revolution in science and technology and the unique productive forces and relations in production that is generated give us reason to predict that this basis of diverse social and economic development versions will grow and grow.

The specific interaction between objective and subjective factors inherent in productive forces and societal development is the other basis of socio-economic alternatives. Essentially different versions of the same socio-economic system (including, for example, the socialist system) can be generated through the existence of differences in the public's awareness and thinking as well as through the interpretation and formulation of objectives within the social and historical process, the specific impacts produced by economic and other trends and directions of human society's evolution as shaped by people's conscious activities. This can result in fairly high uncertainty in social and economic development and, hence, in the freedom to choose versions of main types of socio-economic system. Economic traditions and the national, socio-cultural, socio-psychological, historical and other distinguishing features characteristic of nations also have this result.

The ways in which elements of socialist, or transitional (different from capitalist) relations arise have fundamental effects on alternative modes and versions of societal development. The violent destruction of outdated social structures may generate versions of new socio-economic systems. Other versions may arise when the elements of new systems crop up within a capitalist society evolving as a result of technological revolution and social change: private property becomes integrated with collective property in joint-stock, cooperative and other entities; municipal activity grows in scale: people begin to enjoy better living standards, etc. In any case the contention that socialism may only emerge out of the foundations of violently destroyed capitalism does not appear to be corroborated by history. It stems from an absolutized conflict between socialism and capitalism and from the neglect of the features common to industrial and post-industrial society that were observed earlier in those systems as long-lasting social structures that transcend an individual social system. This clears the way for the rise of elements of a non-capitalist system that are essentially indistinguishable from socialist ones and which keep on interacting with continuing and changing capitalist structures and relations (including certain elements which may arise only if and when capitalist structures undergo changes). As a result, a mixed economy takes shape and begins to operate with its fundamentally different socio-economic modes competing against, and interacting with, one another.

As a matter of fact, any of the factors mentioned above will lead to modifications and versions of general socio-economic systems. A combination of several such factors prepares the ground for a fairly extensive range of alternatives in terms of both structures and modes of socio-economic relations under the general heading of socialism. The same holds true of capitalism and mixed socio-economic systems and their continuously emerging variants.

This is to say that versions and alternatives may be identified and their socio-economic evaluation conducted from the inception of embryonic socialist relations. This also signals the need for a periodic reassessment and adjustment of the guidelines selected and, if necessary, for a complete overhauling of the policy pursued. Lastly, development versions ought to be compared from country to country by drawing upon the experience gained throughout history and by examining the production, technological, economic and social results yielded by various versions as well as man's living, working and evolutionary environment.

1.3. The administrative-centralized system and socialism

It is to be deplored that for a long time the formation of new systems of government occurred almost without any comparison with alternative versions. Moreover such alternatives and relatively independent versions were rejected out of hand.

After the October 1917 revolution a host of domestic and external factors combined to form the Soviet Union's view that a rapid change-over to socialism and communism was feasible by means of nationalizing as much of the national economy as possible, employing numerous methods of coercion, relying on the proletarian state's rule by decree, enhancing the tendency toward social egalitarianism, and by making arrangements for distributing, rather than selling for money, products under tough governmental control. This "War Communism" policy is known soon to have proved a failure, for it paid no attention to our society's real economic, social and other characteristics as well as our economic interests and needs. But over that period a political superstructure and governmental structures were built to consolidate the basic elements of administrative centralization that was the groundwork for transforming the country's social system.

Lenin's ultimate conclusions were that socialism was to be constructed on the basis of economic incentives, regular trading practices and market forces by relying upon cooperatives in harmonizing conflicting interests and integrating masses of people, including small-scale businesses, into the socialist construction effort. This doctrine was corrupted by the War Communism policy but it did help to shape the New Economic Policy (NEP).

The NEP proved a real alternative to previous approaches to instituting socialist ownership and socialism in general. It offered its own alternatives for different sectors of the economy and for the state and cooperative sectors to enter into various interactive arrangements, for the cooperative movement to select from a variety of ways and modes of development and to establish itself as a socialist cooperative movement as well as for choosing diverse ways of market regulation and economic management. What is important is that these and other alternatives made diverse socio-economic relations and modes possible (including ownership relations) and exploitable. But in practice, their potentialities were not put to use. The point was that from its very outset NEP was pursued against a background of bitter conflicts and got entangled in the tendency toward economic nationalization and administrative centralization. Even though that tendency was eroded by the country's changeover to NEP, it was not defeated altogether because it was propped up and fuelled by the political superstructure already in place. The superstructure grew entrenched, thereby reviving and invigorating policies of administrative centralization and economic nationalization. The Soviet economic history of the 1920s demonstrates that immature market arrangements, difficulties in regulating the small-scale business sector, the government's controversial impacts on cooperatives, mismanagement and other similar factors were blamed on the NEP and this was taken advantage of in efforts to nationalize the country's economy. The fact that NEP was abandoned for good in the late 1920s climaxed the process. At the same time, it heralded the ultimate triumph of an administratively centralized society run as if it were a highly regimented state machine that was fundamentally alien to alternative and diverse modes of development. From then onward any alternatives to the established model were viewed as a retreat from the fundamentals of socialism and as efforts to revising its philosophy while the model itself began to be increasingly identified with its substance for no valid reason.

Administrative centralization was made feasible only through to the across-the-board nationalization of property and of economic, social and private life. In turn, nationalization could not be effected outside the framework and methods of administrative centralization. Therefore administrative centralization and nationalization are the two indivisible and fundamental features of the newly formed social system. These features molded other dimensions of this model, in particular, the tendency toward simplified and uniform economic entities: economic ties that used to be motivated by interests gave way to their centralized counterparts based on authority and rule by decree, an administrative command system was instituted to manage the national economy, internal sources and incentives boosting development (that is, making the system self-propelling) were replaced by external inducements based upon the power of the government and the struggle for power that was waged at various levels, while a massive proportion of the workers was prevented from ownership relations and was replaced by a stratum of administration acting on the instructions and in the name of the state. As a result, state property lost or started to lose its feature of public property and from the very outset its designation as such was spurious and deceitful.

The new system proved integrated and fairly sustainable. Its different dimensions shaped one another and were interrelated and well-coordinated. The system was designed to reproduce its own basic properties and qualities.

But what set the system apart was the fact that it was unable to make use of the diverse production and social potentialities inherent in it and it rejected anything that was foreign to the system's fundamental principles. In this sense the integrity of the system was a factor with negative effects on social and economic development, material and intellectual culture and man himself.

The absence of sufficient internal sources and incentives for growth rendered the administratively centralized and nationalized system of relations hopeless in terms of the long-term social and historical process.

In this system workers were prevented from ownership relations and society's property was taken over by the state where political and economic power was monopolized by a few interwoven echelons of administrators and where people's interests, economic management practices and the system's vital functions were dovetailed to reproduce and fortify the system. We argue that this signalled a departure from the fundamentals of socialism as a political philosophy created to serve the workers, to meet their needs and promote the free development of the individual.

At the same time, it should be borne in mind that the administrative and centralized system governing socio-economic relations was built in the wake of a triumphant revolution that sought to transform our society along socialist lines, to abolish the private ownership that had exploited the workers, and to abolish social and class-based differentiation, etc. Some of these features were incorporated in the system of relations that was just emerging. Nationalization as such was claimed to guarantee (though in a somewhat unique way) that no classical division of people into those possessing means of production and those deprived of them would be possible. The system put into effect, in some measure, some other socialist features, namely, the principle of earned income and mandatory work among others. Though often deformed, such socialist features and principles do in fact help to prepare the way for dismantling the administrative and centralized system and provide a variety of modes and versions for the socialist renewal of our society. Departures from the fundamentals of socialism in our approach to ownership relations, power structures, the standing of the individual and freedom of personal development, with the system's low economic and social efficiency when compared to that of advanced capitalist countries, are the reasons why a section of our society has lost confidence in socialist ideas in general, and explains the emergence of arguments for overhauling our socio-economic relations along non-socialist lines.

This demonstrates that the administrative and centralized system did not and could not permit the appearance of alternative versions of socialist development, or of a system of relations that could have been called socialist without reserve and in the full sense. As the system's administrative and centralized basis continued to entrench itself, the more hopeless the situation appeared in countries that had elected to follow this model or some of its modifications.

Accordingly it comes as no surprise now that their change-over to alternative versions of development is painful and that some of those countries face uncertain, fuzzy or even destructive prospects of social change.

1.4. Socialism and a civic society

We can make preliminary conclusions regarding the economic fundamentals of socialist development models in general, and of some particular versions of them. We can do this by considering the basic alternatives to socio-economic development and in particular the multidimensional and diverse processes that affect productive forces and lead to increasingly diversified social structures, by examining the major deviations from socialism made by the social, economic and political model that has claimed to be socialism's standard model, and by reviewing its collapse in numerous socialist countries as well as the experience gained in renewing social relations.

Patterns of socialist development differ from country to country and, hence, the likelihood of these different models establishing themselves under today's

conditions result, by and large, from the nature and scale of the reforms underway in socialist countries. A major factor is whether their administrative and
centralized fundamentals and nationalized social, economic and other relations
in society remained intact, or were obliterated, and how denationalization
proceeded and what results it produced.

In terms of socialism's prospects the scope and nature of denationalization
is of crucial importance. We may pinpoint a few directions those developments
have been observed to take. Firstly, the nationalized sector may have been
reduced to make room for expanding private sector businesses that perform
vital social functions while state ownership remains in existence with changes
going on inside that lead to the formation of its vertical territorial levels,
to the transfer of its numerous functions of appropriating, disposing of and
using resources down to lower levels of management, and to the expansion of
economic independence for state sector production and economic entities.

Secondly, state ownership is converted to common ownership, or joint-
popular ownership, to be more precise, with expansion of diverse forms of
ownership and a more socially responsive management. Thirdly, government-
owned businesses yield way bit by bit to cooperative- and collective-owned
property. Fourthly, the administratively centralized and nationalized economy
breaks down into socio-economic structures of the mixed economy type. If one
of these developments prevails, it may form a backbone for socialism's specific
economic model. These directions for development are not abstractions that
have been conjured up. Some have begun to dominate in individual socialist
countries while most nations witness a real clash of conflicting ideas with no
clear outcome in sight because their reforms and changes are still at their initial
stages.

Various intertwined versions of social, economic and political changes
underway in the countries that had a high degree of nationalization and administrative centralization mirror quite logically their increasingly democratized
social relations. We may continue to witness the numerous combinations of ways
in which nationalized social structures will be replaced by new ones, thereby
enabling socialism to pay attention to and embody the distinctive features of
various countries and their nations' interests and needs as well as to abolish
oversimplification and dogmas in interpreting the system. But the types of
socialist model may be described and identified in terms of one or other basic
socialist criterion, or by a set of criteria defining a given system of relations or
social structures.

The thrust of the changes now affecting socio-economic relations and
structures in the USSR and some socialist countries gives us grounds to conclude
that one of the promising models for socialism to follow may be democratic
development based on real equality, competition and interdependence among
the diverse forms of socialist ownership, social types and economic entities.

Those forms ought to have a dynamic structure that can respond to changes affecting productive forces, a socio-economic environment where resources are exploited efficiently, a society with a socially differentiated structure, and a system of economic interests and needs. A political structure that promotes free and equitable participation on the part of all social groups, that is shaped by the existence of various forms of ownership, by the established form of the social division of labor, and by the social conditions and social environment may serve as a political basis for this model. At the same time, the social, political and professional organization of social strata and groups may vary, for they hinge on a sum total of specific conditions, traditions and levels of democratization and other factors. Denationalized property and other dimensions of social relations are crucial to this model, or, to be more precise, to its socio-economic basis. In this particular case denationalization should be effected both by converting state-owned property into joint property owned in common by teams of workers, with every team member, group of workers and the government representing society in their own way, and by developing independent non-government-owned cooperative, collective and individually earned property with corresponding types of economic management. These lines of denationalization are organically integrated and herald the redefinition of socialism as the basis of as a civic society. When government-owned property is converted into property owned by numerous parties, it is run not by some abstract group but by real people who represent every person and every collective including the state. Relations of ownership become centered on the individual citizen with his interests and needs to be considered while ownership acquires truly nationwide popular features that integrate individual and cooperative (or joint) basic criteria. In contrast with nationalized property, common (or jointly owned) property will not stand in opposition to cooperative and individual property, nor will it jeopardize the latter, for they will have some basic characteristics in common. This is why the conversion of state-owned property into joint property with many owners is a sine qua non for a real diversity of socialist ownership forms to take shape. In turn, various types of cooperative, collective (or joint) and individual ownership erode the basis of nationalized property and encourage efforts to abolish it.

State-owned property with its accompanying political structures proves the most painful and the most important to denationalize. Specific ways and means are likely to vary from case to case but the crux of the exercise is always the formation of multiple ownership of the means of production and profits thereof that incorporate and integrate all the levels and structures of society ranging from the individual to the state. Efforts should be made to identify the way each individual acquires his share of resources and how he contributes to managing the economic and production entities and to social and political relations.

Such relations may be promoted if enterprises with all their assets, resources, land, infrastructure and the like are converted to property that will be owned commonly and jointly by the state, territorial communities of people, teams of workers, groups of people and individual workers, i.e. if the enterprises are run as joint-stock properties owned along internally differentiated lines. It is natural that the promotion of such economic forms should concurrently usher in changes in labor relations, introduce self-management and be followed by more democratic political relations and a reformed structure of political power. Leasehold and joint-stock entities along with share-holding arrangements and matching contributions may become principal ways of converting property along these lines.

The correlations of changes in ownership relations and ways of translating these into reality should, of course, be determined case by case but what is crucial is the desire of everyone who exercises ownership relations to materialize his status as owner and to translate his status into social and political relations and structures.

The emergence of common, or joint, property and leasehold, joint-stock and share-holding arrangements for property administration is crucial to paving the way for the vigorous expansion of cooperative and individual businesses because jointly-owned property already features such basic elements and there is no wall separating these two forms of ownership. In some cases expanding leasehold and joint-stock relations will inevitably result in the conversion of jointly-owned property into cooperative (collectively-owned) or even individual property. And, vice versa, the need to obtain resources, the necessity to meet business interests and other factors may render it imperative to convert collectively-owned enterprises and associations into collective-state-owned property and so on. Social and economic expediency (efficiency), the need to promote integrationary contacts rather than to contrive a scheme for separate existence or rapprochement may become the criteria and incentives for the conversion of modes of ownership.

Several levels of ownership have been established for what used to be state property: national, constituent republican and local (community or municipal). If state ownership is to be renounced a crucial factor in democratizing ownership and making it public nationwide will be the identification of the new levels of ownership. But it should be emphasized that changes in the territorial structure of state ownership must well be coordinated with the conversion of state property into people's property owned by workers as teams and individually. Such coordination will favor, while its lack will threaten, enterprises' economic freedom and to the promotion of a regular market economy in the country. If allowed to happen, territorial isolation will pose a grave danger to the promotion of a market economy and to the transformation of the entire system of ownership. And territorial isolation is likely to take place if teams of workers and individuals are barred from ownership, or if teams of workers and individuals are not confronted with the need to compete for markets, including local, interregional and national ones. If territorial levels of state ownership are established without changes in the nature of this ownership, this will be fraught with the threat of further bureaucratizing of ownership relations. The existing administrative-centralized system may encompass local levels. And, as rule, local bureaucrats are more dangerous than those in the central government.

Thus, efforts to denationalize are attempts to erode the state's right to ownership. The state continues to wield its power as an economic center that relies on economic inducements for regulating production and, hence, appropriates, exploits and disposes of its allocation of the means and results of economic endeavors. The crux of denationalization is instituting in practice rather than in words multiple ownership, such owners ranging from a specific individual through a team of workers to the state as representing the entire society. The point is that the state should cease to be the sole owner and emerge as one of several owners. Such a conversion will also become central to the efforts at dismantling the administrative–centralized system and at building internal sources and incentives for socialist development.

State ownership as such will also take its proper place among diverse socialist economic structures. Denationalization and the establishment of common (or jointly-owned) property cannot replace state ownership as a distinct form of ownership just as nationalization is not identical with the state or the economic role it plays. Nationalization is state totalitarianism that corrupts not only socialist but also socio-economic and political relations. Nationalization is the suppression by the state of all forms and types of ownership, except for state ownership, and converting the latter into a tool of state diktat over the economy and the people. Denationalization allows the people to regain their status as owner alongside other forms of state, cooperative and individual ownership. State ownership will be an independent phenomenon that society needs to promote certain promising and experimental production facilities, to elaborate research and technological programmes, and to stockpile contingency reserves among other things. Such state, not nationalized, ownership abandons its role as the only permitted form of ownership. It will expand within an overall and continually evolving system of socialist ownership. It may operate in joint-stock and lease-holding arrangements and change into joint ownership, or vice versa. In some cases jointly-owned or collective enterprises may be taken over by the state as a result of an agreement, or through purchase.

With expansion of the number of possible forms of ownership, there comes a greater diversity of forms of civic organization and this should provide a social and economic ground for private property to be exploited in the interests of socialism and within clearly delineated boundaries. The modifications that were introduced in the Soviet Union's socio-economic relations have however not achieved this, as unresolved theoretical difficulties were merely transferred from the academic sphere to that of practical application and this brought about diametrically opposite and unsound proposals. One such approach views private property as a would-be structural element of the socialist system itself. This erodes the social components that make socialism what it is, namely, a system that makes everybody an owner of the means and results of production. The opposite approach relies on the latter argument to reject altogether the admissibility of private property under socialism. Certainly there are elements of private ownership within a socialist system and claiming that such relations are impossible, or inadmissible does not render them non-existent in reality if popular and economic needs make them necessary. The available resources and the nature of some production and economic activities set the stage for private businesses. If an outright ban is placed on them, they resurface as clandestine and illegal operations that have grave negative effects. It appears better for the state to allow and regulate private property and private business operations that are essentially unsocialist, to provide guarantees of social protection for privately employed personnel, and to place fair ceilings on the profits made as a result of such business operations. That such an approach is sound has been demonstrated in several countries. At the same time, it should be borne in mind that private businesses will evolve and maintain very close and organic relations with various publicly-owned entities that will make a substantial impact upon those businesses' philosophy and social responsiveness.

The existence of diverse forms of ownership including, above all, state ownership becoming converted into common (joint) ownership, with the massive expansion of cooperative and other forms of ownership introduce fundamental changes to the role commodity and market relations play. Commodity relations and market forces are alien to, and destroy the foundations of, a nationalized administrative centralized system. Commodity relations and market forces come naturally to property that is in multiple ownership and to a diversity of other ownership forms. The "problem" of the market economy ceases to be a subject for people to debate and regains all of its fundamental aspects and dimensions.

A diversity of forms of ownership affects the entire system of social and economic relations by transforming relations in employment into relations of membership within teams of workers, cooperatives, etc.; economic links established by fiat are changed into mutually beneficial contractual arrangements; directive planning practices that lead to huge imbalances into realistic planning based upon economic inducements and regulated incentives. Further, such diversity helps in distributing basic necessities and a part of the surplus production that comes into the worker's possession as his or her private share or dividend.

The development of credible social guarantees and social protection for the worker emerges as a crucial dimension of a new system of relations, for social differentiation and its controversial process of development will be recognized rather than glossed over as one of the array of driving forces and motives that propel societal evolution. In this particular case alongside a host of social guarantees that are offered by the state in its capacity as society's representative there arises the need for a network of non-state guarantees that will be provided by the payments of the workers themselves as well as by their teams, groups and associations.

A network of social guarantees and a system of social protection supported by teams of workers and cooperatives must become one of the dimensions of social and economic independence. Cooperative arrangements entered into by the population may offer particularly great and diverse prospects in this respect. Cooperation, or, to be more precise, a system of government resting on civilized cooperating workers is not merely a technique for conducting production and economic operations.

This is also a method of organizing the social activity of the working people. The reviewed bases of production with their economic, social and political consequences form in their unity a common basis of civil society as a social mechanism which provides a double protection for each individual, collective, territorial association and other community of people: first, their protection by the state acting as a representative of society, and, second, protection from the state. The basis for the latter kind of protection is provided by the preconditions for economic, social and political independence of every individual, or collective. It is precisely this double protection which will be a guarantee for free development of the individual and, as a result, of free development of all.

1.5. Cooperative socialism and mixed economy systems

One of the options for the development of socialism might be its establishment primarily along cooperative (collective) lines in which case the majority of production and economic units and areas will have to be involved. Some early indications of this can be seen in the ever increasing orientation in our country toward enterprises becoming the collective property of the working people, in the slogan "factories to the workers", and in the new interpretation of leasing and other state and collective forms as temporary and insufficient. In speaking about this option one should note that the past criticism of cooperative socialism resulted from the impossibility of a purely cooperative socialist system due to the limitations of cooperative ownership by groups of people, and from the inevitability of such relations among them as are determined by the fact that they have different rights of ownership and consequent differentiation in their economic status. Indeed, such pure cooperative socialism is impossible, as indeed any other pure society is impossible. The world development experience shows that capitalism inevitably forms state, or public resources to regulate the economy, to resolve social problems and for other purposes. It is natural that a cooperative-type economy, especially when cooperation is both horizontal and vertical, provides a significantly greater basis for creating social means that are necessary for the appropriate regulation of inter-cooperative relations.

Further, denying any possibility for socialism to develop on the basis of cooperation used to be taken together with the interpretation of cooperation which saw in it a transitional socio-economic form which was not consistently socialist. Nevertheless, such an interpretation was based on there being a restricted range of forms of horizontal cooperation (with a multitude of individual cooperatives), without taking into account the possibilities and requirements of vertical cooperation. Consequently, the interpretation was based on an extremely truncated interpretation of what social ties could be produced by cooperation, and under current conditions that is not right. The development of vertical ties, based on the principles of cooperation, can make it possible to integrate various economic levels, provided the cooperative basis of lower-echelon cooperative production and economic units is retained. That is why it would be wrong to renounce in principle the possibility of developing socialism primarily on the basis of cooperation with a sufficiently complex system of relations.

If socialism is created primarily on a cooperative basis, significant peculiarities will be introduced into the system of socio-economic relations and into the trends of their development. These peculiarities will cover the market, distributive relations and the processes of growth as a whole, as well as the dynamics of social differentiation. Accordingly, the system of social guarantees will have to be specific. Nevertheless, given the fact that there will be no cooperative socialism and that state, or a mixed-type economy will become necessary in some form or other, the given model will be relatively close to the preceding one.

It should also be noted that this option may only become possible when we have a number of specific conditions in place, that would include among other things the cooperative economy having acquired its own traditions, a smoothly operating mechanism of horizontal and vertical cooperative ties, a broad popular consensus as regards the benefits of cooperation, and when we have a high degree of civilization both within the cooperative sector, and in society at large.

One should not also reject as an option for developing socialism a mixed-economy model that has emerged and is functioning in a number of highly developed countries as a result of implementing what are primarily social-democratic programmes. Simplified interpretation of social systems in these countries as being state-capitalist, or simply capitalist, is not substantiated because there we have complex socio-economic relations and structures, sufficiently powerful state, municipal and cooperative sectors and systems functioning to a significant degree in the interests of the working people, with restricted opportunities for private capital (including the remittance of a large share of profits, strict labor legislation, strong trade unions, etc.), and developed forms of cooperative protection of the working people as consumers. These are all elements of the previously reformed and non-capitalist economy. Obviously it would be an oversimplification to regard this system as being defined by a sufficiently powerful private capital sector. In this case we are talking about something else – a possible emergence of socialist, or socialist-oriented, relations on the basis of major structural and social reforms, the processes of introducing joint-stock ownership and democratization of capital, creation of state and cooperative sectors in the economy, a better standard of living for the working people. These relations should be consolidated, become competitive and interact with the remaining and changing capitalist structures given gradual in changes in the social system (though such changes are not obligatory, and far from being inevitable). It is important to note that a mixed economy is not a mixture of socially different elements, but a sufficiently solid unity, capable of reproducing itself, and managing sustained development both in terms of production and of social development, in parallel with concurrent transformation of socio-economic relations and structures. It is precisely this feature that makes it possible to talk about the mixed economy being a model for transforming capitalism and changing society on the basis of socialism through democratic means while respecting the interests of all the strata of the society while the appropriate requirements and preconditions for such a transformation are being created. Interaction within a mixed economy system is an additional and rather effective source of their dynamism, socio-economic efficiency and democratic nature.

Speaking about mixed-economy models, as well as about the options of developing socialism mainly along cooperative lines, one should note that all constitute civilian organizational forms. In other words, it would seem that the development of socialism as a civilian society should become one of its characteristic features.

2. State Ownership: The Essence of Changes

2.1. Joint-stock ownership

Joint-stock ownership is one of the major conditions for making state ownership nationwide, i.e. for making state enterprises the common property of all working people, of each working collective and of the state. That is why introduction of joint-stock ownership should influence the relations within an enterprise, between the state and enterprises, and among enterprises themselves. In this connection a major question is that concerning the participation of the state itself in this type of ownership. Real changes in the principles of state ownership are possible provided the state is one of the partners in a joint-stock type relationship, participating through owning capital, as well as directly through becoming a share-holder, buying and selling the shares of its own and of other of joint-stock societies, companies and enterprises. In such a case the pattern of ownership changes, and great possibilities are opened up for economic regulation and provision of incentives for production.

It would be inappropriate to equate joint-stock relationship with other means of attracting additional monetary resources. This relationship is more than just another way of resolving financial problems.

It becomes apparent that there are many ways to develop joint-stock relations, and one should be aware of a choice. First of all, joint-stock relations may exist wholly within state enterprises, or cooperatives, with only the workers of a given unit or enterprise participating in them with a view to resolving matters of production and social issues of interest to its collective. These relations have now become rather widely, though in a limited way, practiced compared with other forms of joint-stock relations.

Secondly they can transcend a given economic unit. There exists already a need for joint-stock companies, societies, and associations in the production sphere, in that of social infrastructure and in other fields.

Besides, a great number of very serious social problems can be resolved by developing entirely voluntary joint-stock associations with the participation of interested enterprises and individuals: for example, such matters as recreation facilities, communal services, catering to everyday needs, increasing production of those goods which are in short supply, etc., can be run on this basis.

Joint-stock relations can be developed within individual administrative and territorial units and regions. Participation as share-holders of both individuals and enterprises can become a major inducement to developing regional economic accounting and resolving production and social problems that may have a territorial basis.

Joint-stock relations cannot be considered simply as issuing shares to individuals, and to the workers at individual enterprises. We are talking here about a much broader complex of relations which could permeate the entire economy and various spheres of social life, and it seems that this is the angle from which the prospects for developing joint-stock relations should be approached.

The attitude toward joint-stock relations in the national economy is ambiguous. In particular, there is a perception of a possible danger, for under these conditions there is room for unearned income, and the profits received for past work, and, consequently, a way is thus cleared for the violation of one of the most important principles of socialism – distribution according to work done. There have always been various forms of unearned income – inheritance, getting interest payments for the money kept in a saving account, and a whole range of other forms. In this particular case unearned income as such is not important, but of importance is the nature of this income. Indeed, income in the full sense of this word cannot be considered unearned, for in this case investment is made out of those means that are received as a result of one's own personal work, and the investment is made to develop the national economy and to raise the general standard of living. As a result of such investment people postpone the satisfaction of their immediate needs for a certain period of time, sometimes for a very protracted period of time.

The entire society stands to gain from this, and should people get some income as a result, there is and there can be nothing in it that contradicts the basis of socialism. On the contrary, making use of joint-stock relations will really help to motivate people to contribute to the development of the common economy, for they become real participants in this common economic activity.

Under favorable conditions, i.e. provided economic reforms are implemented, and all areas of social life are restructured, one could, in the final analysis, have various forms of joint-stock relations, enterprises which would function on the basis of pooling the capital of both the workers and the enterprises themselves, etc. One should aim at creating exactly such a system, compatible with a well-developed system of leaseholds, cooperatives, and an entire range of other new forms. This would allow a real restructuring of person-to-person relations during the process of acquiring both the means and results of work, thus making people the real owners within the framework of public ownership, which would no longer be in opposition to the individual, while the individual himself would become inseparable from this property.

2.2. Lease-holding relations: problems at the formative stage

Leasing has been accepted as a major area for updating economic relations, and imparting to them inner dynamism. The Law on leasing relations has now been passed, providing both the order and conditions for their development.

There are few enterprises that are indeed working on the basis of leasing, and the liberalization of leasing law has not produced much effect. It comes as no surprise, therefore, that direct attacks on leasing have now begun. At the first Congress of USSR People's Deputies there were demands to make public the names of those who "invented" cooperation, leasing and other new forms of economic activity, evidently the intention being to have a heart-to-heart talk with them. Such demands are not feasible, for life itself gives rise to leasing. But the same life gives rise to considerable forces ranged against this phenomenon. The ways and means of resolving the acute economic and social contradictions related to leasing depend to a large degree on how well the philosophy and prospects of the renewal of socialism are understood. The latter in their turn are largely determined by the fate of leasing and other new forms of economic activity, which could provide the means for a real change in existing ownership patterns, as well as other elements of the socio-economic system.

Both the prospects and difficulties of developing leasing become clearer if we perceive the need for it in a socialist economy, the functions it is supposed to perform there and its place within such a system of economic relations. To do this, it may be necessary, first of all, to explain why until recently leasing as a form of managing the economy had not existed, and why the very idea of leasing seemed bizarre. For a long time the entire system of economic relations has been based on direct state regulation through all phases, stages, sectors, production and territorial units, etc. Under these conditions the working people, working collectives and groups could in reality dispose only of their wages, in ways which also were strictly set by the state and were limited to include only part of the profit. In this sense private ownership of the income derived from work was of a secondary nature. Specific people and collectives have not become owners of the means and fruits of work. Under these conditions leasing relations become impossible, for leasing involves a transfer of means to appropriate collectives, or individual workers for their use with all the resultant possibilities of acquiring and disposing of both the products produced and the income earned.

In other words, leasing would not allow across-the-board state ownership of the means and fruits of work, or at least would restrict this ownership. The argument against leasing at that time, and sometimes even now, was that under socialism everyone is an owner, and that is why it is impossible to have lease-holders and those who give leases. This argument was purely apologetic in character, for under the real system not everyone was an owner. The real reason why it was excluded from production and economic relations was exactly the desire to preclude the possibility of having everyone become an owner.

At the basis of state ownership was the process of raising to the absolute level certain trends whereby production and its social forms were made public property, and, above all, the trends toward centralization and direct regulation. Such a process patently contradicted the variety and multidimensional character of the processes of making production really public, and this in practice lead to their being restricted and checked, while the formal, administrative and centralist character of this process came to prevail over its genuine form.

The primacy of administrative, centralist and state relations could not remove the existing and complex socio-economic structure with its strata, groups, collectives and different interests, for this structure is determined by all production forces, as well as characterizing their status. The preponderance of state ownership has been and continues to be in direct contradiction to a differentiated system of economic interests, thus sharply restricting social and economic development. This results in a search for possible tools to reconcile the interests that have become isolated, and this happens even within the administrative, centralist and state system of managing the economy. One such major tool was economic accounting, which made it possible to manage the state-type economy under conditions where some economic independence had been provided to enterprises and other production and economic units. The particular feature of accounting consists of independence (the forms and limits of which could vary) being allowed within the framework of state ownership. That is why cost-accounting independence is, to a significant degree, only of conditional or formal character and does not resolve the problem of how the state economic system acquires sufficient inner resources and incentives for self-development.

Such incentives are provided by new forms of economic management aimed at doing away with the state-dominated character of the economy, among them leasing. The possibility and necessity of leasing is determined by the fact that property relations are characterized by a multiplicity of levels and subjects. Leasing itself provides practical ways and means to differentiate of the processes of acquiring and disposing of the available means within the confines of public ownership.

First, it is through a leasing agreement that production and economic functions, obligations, responsibility and relationship among the participants in leasing relations are determined, divided, or united. Thus an economic foundation is created for inner differentiation of the processes of acquiring and disposing of the available means, while ensuring their unity.

Second, leasing provides a basis for acquiring ownership of the products produced by lease-holders, the leasing profits, and of the means acquired by using this income. As a result, leasing provides the means to do away with state-type ownership, while making the working people the real owners, on a par with the state.

Third, leasing payments make it possible to redeem and increase the value of the leased public facility as well as to enable further use with regulation of its scale and consolidation of its social character.

Fourth, due to the development of internal leasing relations, especially coupled with the principles of share-based distribution of part of the leasing profit among the members of leasing collectives, each worker can become an owner.

Fifth, the basic elements of leasing relations (leasing agreement, leasing payment, leasing profit) permit the combination of various economic interests through providing the framework for means to be acquired.

All these aspects characterize the main functions of leasing and its place within the system of economic relations as another means of realizing the multi-subject character of property relations and ensuring its popular and nationwide character. On the whole leasing relations ensure redesignation of all those involved in property as owners, and their socio-economic interaction makes for a social (nationwide) character of ownership.

The role of leasing as a means of transforming state-type ownership into nationwide testifies to the possibility and appropriateness of making leasing apply to various spheres and sectors of the economy. At the same time extending leasing to cover major enterprises is most important from the point of view of transforming socio-economic relations. Attempts at restricting leasing mainly to small enterprises, small collectives, and internal economic relations in fact contribute to the preservation of the supremacy of state ownership, and, thus, the entire administrative and centralist system of relations. The resilient nature of this system, its stability and incompatibility with leasing, together with ingrained conceptions regarding the permanent and dominant nature of state ownership explain that leasing is being considered as equal to the cost-accounting system, to the existing semi-contractual and similar economic forms. Leasing, indeed, serves to make more profound the basic elements of cost-accounting, while making them real. But this is an entirely new phenomenon. The line which separates leasing from cost-accounting is the emergence of the ownership of collectives and workers of the means and results of work, of the profits derived, and, consequently, the opportunity freely to dispose of them and use them in accordance with the needs and interests of lease-holders. Leasing is present where there are real changes in property relationships in the direction of making them non-state in character. Unless this line is crossed, there can exist forms of cost-accounting that resemble leasing, or forms that replace leasing, but there can be no leasing as such. This should be taken into account, for there are multiple forms that are wrongly united within the notion of "leasing", and should be kept in mind in assessing the efficiency of such "leasing".

Speaking about the economic systems that make use of certain elements of leasing, and which have features in common with leasing, it would seem that one should take into account that their emergence (already in varied forms) is inevitable, for the transformation of state-type ownership into nationwide, including various forms of popular ownership, is a complex process requiring several transitional stages. And the line itself beyond which collectives and workers begin really to participate in the acquisition of the means and results of work can be rather conditional. The new relations of leasing and ownership often emerge before they get legal shape. It is important to appreciate the transitional character of various semi-leasing forms, their instability and incompleteness while having a programme of making more definite the basics of leasing, and ensuring genuine leasing relations. This is a normal way to develop leasing.

It is another thing when because of expediency or other considerations leasing is equated with similar forms that are restricted to collectives, and when there are no prospects for its development. Thus begins the repudiation of leasing, making hopes for real change in socio-economic relations futile. Such trends of "developing" leasing are not accidental and they are extremely dangerous, for they spring from the incompatibility of leasing with the administrative and centralist system, from the latter's intention to deprive leasing of its democratic content through linkage with development of people's ownership.

The main obstacle to wide development of leasing and other new and related economic systems is the state administrative and centralist system. It means that in order to develop leasing and new forms of social relations generally one should reduce the immune qualities of this system. Common sense, supported by the rich experience of recent years, dictates that it is impossible to change the system of socio-economic relations by modernizing only some elements of the economic mechanism, or by implanting into this system certain new socio-economic and economic forms alien to it. In particular, the current situation as regards leasing and cooperation testifies to this. Cooperation has found itself in a situation where it is no match for what is in fact the non-reformed state economy. On the one hand, cooperative economy, unless it is integrated with the state economy, is deprived of normal sources of material and technical procurement. On the other hand, it enjoys broader opportunities for manoeuvre, responding to the demands of the moment, using the available means, and the profit fund of workers.

Instead of the emergence of relations of partnership, and the normal and necessary competitiveness of state and cooperative enterprises we now see not only economic but also social opposition related to the differentiation of profits. There exist several trends undermining cooperation: there is growing administrative and economic pressure on it, and some cooperatives increasingly try to obtain immediate profits, while there exists speculation, ties with the black economy, etc.

As regards leasing as currently practiced it is generally subject only to cost-accounting and cannot become genuine leasing, and this is shown by the restrictions in the disposal of the products and of the profits received. Further the lease-holders often depend on those who grant leases (agencies) for resources and facilities.

The attempts at rapid introduction everywhere of internal economic leasing in collective and state farms (the so-called lease-holding) have revealed the lack of economic and social preparation for the extensive development of this form of economy management and have even come to discredit the leasing idea and prevent development of leasing relations. This is not surprising, for a real transition to leasing relations on a broad scale inside sectors is impossible because these sectors operate in conditions which are far from those required in leasing. There are frequent cases of direct administrative interference in the affairs of the leasing work teams, breaches of obligations by the lease grantor, etc.

Along with positive development in terms of higher labor productivity, low demand for extra workers, and expanded production, the first results of the introduction of leasing also point to occasional instances of stagnation and even reduction of production volumes. This is because where leasing relations are sporadic in nature, where there is no competition, and where profit-making opportunities are limited, a mere reduction of workers and material resource savings (obtained by reducing the production volume) may turn out to be more beneficial to each member of a leasing work team than an increase in the number of workers and the concomitant expense of increasing production.

In all probability a rational way to weaken the capacity of a nationalized system to reproduce itself and to help new economic relations, including the establishment of new types of management, would be to impose certain conditions. The most important would be a step-by-step but mandatory and comprehensive alteration of the entire economic machinery including planning, material and technical supplies, financial and credit relations, pricing, methods of forming an organizational and economic structure, the character of implementing economic links, etc., with a view to developing market contractual relations on the basis of the principles of mutual benefit. Only such far-reaching changes in economic relations will permit transformation of the existing forms of management to enable introduction of new economic activities. It is precisely the lack of a comprehensive approach to economic reform, with a step-by-step reform of various aspects of the economic machinery, that has actually brought reform to a standstill, injected inconsistency and all kinds of distortions to the development of leasing, cooperation and other new types of economic relations, with a negative effect on economic growth and development.

Secondly, there is the need to ensure interconnection and proportionality of change in the economic machinery and to introduce new mechanisms. It does not make sense – or to be more precise it is even harmful – for growth to outstrip the emergence of new elements in the economic machinery because then these elements are destroyed and negative trends are generated. But if the development of leasing, cooperation, etc. proceeds fast under pressure from below, then it is necessary to accelerate radical changes in all the aspects of economic relations and above all in the mechanism that operates the economy.

Thirdly, it is important to develop various forms of leasing and especially those which encompass several economic levels. Depending on a specific situation, it is necessary to develop collective and individual leasing, multitiered and single-level leasing, etc. The forms and terms of leasing should be distinguished according to different branches and spheres of the national economy, according to types of enterprises and other production-economic units. At present the combination of intra-economic leasing and leasing at the level of an enterprise as a whole is especially important, provided of course that the principle of complete freedom of choice in the development of both is applied. Any attempts at setting up extensively intra-economic leasing work teams without changing the whole status of an enterprise are utopian. It is necessary to reform the economic and social status of an enterprise and its work team by granting to the latter the rights of a collective lease-holder with rights and responsibility ensuing therefrom. In turn, the leasing of an enterprise is not feasible without a radical revision of intra-economic relations, including relations involving the use of resources (leasing them to the respective units, groups of workers or individuals depending on specific conditions) and labor relations. The latter should be changed in principle. First of all, a collective lease is incompatible with the setting up of a work team on the basis of hiring labor. A leasing work team is a union of equal workers who have united themselves in a partnership and who cannot hire each other. Consequently the right of membership of a leasing work team should be introduced instead of hiring, and the rights and duties of such a member should be on the basis that each has a share of the profits and resources acquired and of responsibility for the overall result. Accordingly, it would be well to establish a procedure for admitting members of a leasing work team to partnership and for terminating the partnership if necessary. Fourthly, it is necessary to establish interrelationships in the development of all aspects of new forms of economy management and above all those that involve leasing, joint-stock and cooperative forms. The consistent development of leasing involves the selling of shares by each owner as a condition for giving effect to the status of owner. The intrinsic presence of share-based or joint-stock principles in multi-owned property is the basis for the development of all kinds of share-holding entities which vary according to their functions, composition, scope and areas of activities.

There are many ways in which leasing can be transformed to share-holding and it cannot develop further by any other route. It is impossible to allocate the means and results of labor by each of the owners or to determine the measure for such allocation without share-based or joint-stock principles. But that is not the only point. The allocation of a profit derived from leasing and its use can be effective only if there is room for economic manoeuvering with respect to this profit or if it is possible to keep track of the existing market and the trends stemming from it, and this involves the concentration of resources, their multisectoral and inter-regional transfer or their use in terms of assimilating new technologies and other factors which characterize the dynamism of economic activity. In the circumstances where work teams and individual workers become co-owners of resources with the state, the establishment of joint-stock enterprises, companies or societies and other entities becomes a method of ensuring the mobility of resources and increasing the internal dynamism of an economic system.

The participation of the state as a partner through investing funds in the development of share-holding entities and lease-based enterprises would be a practical argument to refute the misgiving that leasing and share-holding will lead to the break-up of the public property into collective and private. It would be an evidence of the changes in our economic system that would mean among other things the establishment of truly common people's property which has not existed until now. Leasing and share-based relations are inseparable from each other in dealing with this global task.

Leasing and especially share-based relations have a number of features in common with cooperation. For that reason their development is an economic basis for a real achievement of comparable conditions for the activities of enterprises and farms. Accordingly, it acts as a major guarantee for the stability of the process of development of cooperation.

The emergence and implementation of the ownership of lease-derived profits by lease-holding work teams and the production and social funds obtained with these profits will virtually destroy the wall which separates the present state-run economy from a cooperative economy. The work teams' ownership of lease-derived profits and production funds obtained with them offer the possibility for the work teams to buy out enterprises as their property or to transform them gradually into such property as the leased funds and their reproduction by using lease-derived profits depreciate (in case the state fails to make supplementary investments to a given enterprise). Even now there are work teams which are willing to buy out their enterprises. In this connection views are expressed regarding possible negative implications of this process for socialism (here again they see common peoples' property replaced with cooperative property). It would seem that such misgivings are groundless. To begin with, common people's property does not yet exist, it will have to be created by transforming the state property that has become a substitute for it. Furthermore, it should be understood that in the course of establishing and expanding leasing relations, bilateral processes are likely to occur: one will develop in the direction of turning a portion of lease-based enterprises into cooperative entities while the other will – as cooperative enterprises become joint-stock entities and as state funds are leased to them and their production activities become integrated with intermediate enterprises – actually turn purely cooperative entities into joint collective and state entities. A free transition of a portion of enterprises from one form of ownership to the other is a normal process of the dynamics involved in ownership forms, a process which is conditioned by economic expediency (effectiveness).

Besides, the problem of buying-out – in contrast to the problem of developing the leasing system in general – may affect relatively small enterprises for which the cooperative form can be more reasonable. As to large enterprises in the fund-generating sectors, they cannot be bought out by virtue of the fact that the state is in a position to invest capital there or simply because of the vast expenditure that would be required to buy them out.

Generally speaking, cooperation will not absorb leasing. On the contrary, it creates a number of premises for the development of lease-based relations because cooperatives are already the owners of their profits and can partially

use them to lease the necessary resources unless obstacles of a different kind
are raised.

3. Cooperation: Tactical Setbacks and Strategic Prospects

It is difficult now to find anything in our economic and social life arousing
more passion than the matter of cooperation. Indeed, over the last two or
three years conflicting and at times even extreme concepts and proposals have
emerged: from the desire for a general and complete transition to cooperative
forms to their no less complete repudiation. But extremes are born, as a rule, of
strictly local circumstances.

Actually, as regards cooperation there are two current approaches. One
is based on the tasks of the real restructuring of the entire existing system of
social and economic relations in the country, including patterns of ownership;
on the immanency of socialism's diverse social forms of economy, including
cooperation, and the need for their interaction on the basis of socio-economic
equality and competitiveness; on extensive assimilation of cooperative principles
by the state economy – which may become one of the principal arenas of
its transformation into an economy of all the people; on the development of
cooperation in various spheres of production and economic activities; on a
gradual involvement of the entire population of the country in some form or
other of cooperation with a view to satisfying more fully each person's differing
requirements; and on the implementation of Lenin's conclusion about socialism
as a system of civilized cooperators.

The other approach reflects in fact the interpretation of cooperation as
a form of economy of restricted applications which deals primarily with some
current economic and social tasks related to the play of market forces, social
differentiation, the growth of income outstripping the work contribution, and
some other socially negative phenomena. Accordingly, the redistribution of
labor and material resources between state and cooperative economies to the
benefit of the latter appears to be irrational. By the same token it would seem to
be unjustified for cooperatives to have the broad economic freedom to choose
their spheres of activity, centralization, sales, etc., which enables them to be more
flexible than state enterprises of the same profile. This approach is implemented
in practice through various decisions, instructions and circular letters which
forbid or restrict the use of cooperative forms in a number of activities, limit
to negligible amounts the cash that cooperatives get at any one time, and so on
and so forth.

Sometimes the elements of both approaches become intertwined and
adjacent, as, for example, in the decree on cooperative taxation, which makes
the fairly broad opportunities for cooperatives to get tax benefits contingent

on the way local authorities treat them, injecting thereby instability and flux in the conditions of the development of cooperation, resulting in a situation where cooperators have a sense of fragility of their position.

These approaches and their respective status in the process of implementing cooperative development are stipulated by several factors. They are affected, of course, by clashes of views with regard to understanding the philosophy and fundamental features of socialism and the renewal of our society. Cooperation fits poorly into the central-administrative system of directly assigned tasks and regulations. In a system where there is planned development and where the exclusive priority of state ownership prevails, where there are no alternatives to socio-economic processes, etc., cooperation is not favored.

Cooperation becomes an absolutely essential and equal form in the concept of democratic socialism, which relies on awareness of the complexity and diversity of real socialization, the development of production and other social links, and the consequent need for a diversity of socio-economic structures, forms of ownership and types of economy, the possibilities for combining them in different ways and the resulting alternative options for the development of society. This concept is based on the proposition that social and economic forms must be devised to confer the status of owner appropriate to labor activities, on each specific person, work team, association, social stratum and group of people, and on society as a whole. And this, in particular, presupposes the implementation of economic links and their regulation on the basis of the principle of the mutually beneficial combination of interests of all the participants in economic relations, which is required, in turn, by the principles of economic democracy, self-management, self-financing, market development, etc. The implementation of such principles implies both the practical use of cooperative principles throughout the public economy and the creation of premises and conditions for the development of various forms of cooperation.

But the chief factor which stipulates fundamentally different attitudes toward cooperation is life itself and the economic and social contradictions that exist. Different theoretical concepts of socialism and its correlation with cooperation are in fact merely the expression of different, at times conflicting understandings of the nature of life and its contradictions.

It so happened that cooperation (or to be more precise new kinds of it) was the first major social and economic structure born of restructuring social relations in the country and operating on the basis of truly new economic principles. New cooperation began to take hold at a time when in the sphere of state economy and state-owned cooperative forms changes in some aspects of activity were still incipient with the old principles remaining virtually intact. The new cooperative principles are associated with development of businesses at their own expense (i.e. self-financing and the pay-your-own-way principle), with relations of competitiveness in the spheres of production, sales, involvement of material and labor resources and the user, manoeuvrability, entrepreneurship, quick reactions to market conditions, etc. As a result they have come into conflict with the central-administrative system from which all this is almost or entirely absent. Most notably absent are competitiveness, manoeuvrability and entrepreneurship. The market for the means of production does not exist while the market of goods exists largely conditionally. But then everything is preplanned, namely who should get what, where and how much – from material resources to wages and salaries – and who must pay what to whom, where and how much. One of the principles of this particular system is that you cannot get more, be it resources or wages; another principle is that you should not give more than what is prescribed, otherwise economic troubles will result. To combine cooperation with such principles and conditions would be to make it impossible for people to be involved in cooperation.

Therefore the mounting contradictions between cooperation and the prevailing social and economic conditions, which had a negative impact both on the nature and purposes of cooperative activities, were inevitable. Cooperation has grown to fill the free economic niches which were created by the central-administrative system and which this system is unable to fill. It turned out that there were quite a few niches in terms of unsatisfied needs of the people for various goods and services, in terms of needs of the state enterprises themselves involved in performing certain kinds of work, scientific and technical maintenance, etc. Initially unaffected were the sources of material resources (secondary raw materials, the retail trade, especially goods that were not in demand, etc.). However, the resource problem has been the most difficult one for cooperation from the very outset.

The beneficial taxation regime for cooperatives, which remitted three percent of their growth income to the budget in the first two years with the subsequent increase in the tax up to ten percent maximum, has had a very beneficial effect on the development of cooperation.

But even the first signs of expanding cooperation, the emergence of increasingly new spheres of cooperative activities, with cooperatives evolving from work teams into small-size but well organized enterprises with a well established rhythm of operation, have sharply aggravated the problem of material and labor resources for cooperation. The opportunities – which have by now been revealed – to earn in cooperation more than in state enterprises under the existing system of wages and regulation of their activities, have been instrumental in finding a quick solution to the labor and resource problem of cooperatives. But the material and resource insufficiency has become even more acute. Cooperation has increased the purchases of some goods from the retail network thereby causing anger among a significant segment of the population.

The vital importance of material and technical supplies, the high wholesale prices of products sold to cooperatives unless they work to fulfil a state-placed order, the inevitable need to get hold of various materials by paying over and above their price, etc., resulted in high prices for products made by cooperatives. They were also determined by shortages, by the existing correlation between supply and demand, and by the lack of incentive on the part of state enterprises and organizations to compete with cooperatives by marketing of goods and services at lower prices.

As a result, there emerged and began to grow popular discontent over cooperative prices, and over cooperation itself. This discontent has become particularly wide-spread because of the differentiation in wages in cooperative and state economies. But they fail to take into account the reasons for the relatively high incomes in cooperation, which are due above all to its economic independence and the incentive in the results of its activities, as well as the reasons for considerably smaller wages in the state economy which are due to its administrative over-regulation, non-differentiation and rigidity of the wages system which is divorced from the real results of activities, etc. What is important now is the very fact that wages differ. A paradoxical situation has emerged: discontent over the central-administrative system which is incapable of meeting the requirements of people in food, goods, services, etc. or of providing an opportunity to working people to earn high incomes and achieve a good living standard, has spread to include cooperation which itself weakens bureaucracy.

Speaking of the high income of some cooperators and the high prices for cooperative output and other aspects that give rise to social strains, one should, of course, take into account the entire set of circumstances that contribute. Thus the growth of income in cooperatives is partially related to the fact that the bulk of the income obtained is used to pay the workers while a relatively small amount is invested into production and creation of the material and technical base for cooperation. But then there is not much room for investment because you cannot get hold of equipment, hardware, material, that is to say the growth of cooperative income is stimulated by the fact that there is no smooth functioning system of material and technical supplies to cooperatives.

At the same time the lack of adequate stability in implementing the cooperation development strategy and the insufficient prestige of cooperation per se have intensified the inflow of people to cooperations who are after big and relatively easy profits. Hence the mono-directional use by them of the market condition, scarcities, etc. Such instances are not isolated and, although the state sector is beset by very similar problems, solving them is particularly important for the cooperative sector. The proliferation of the money-grubbing philosophy is the result of the underdevelopment of cooperation itself and the conditions of its activities. But these phenomena also distort cooperation from the inside. The principal method of freeing cooperation from "uncivilized" fellow-travellers, on the one hand is to move it into the phase of mass scale development, while on the other hand, is to speed up the economic reform and restructure the conditions of state economic activities. Then competitiveness will make cooperation an unprofitable sphere for the application of the money-grubbing idea. These are also the main ways of overcoming social distrust of cooperation, and of enhancing competitiveness between state and cooperative businesses from the standpoint of opportunities for improving the living standards of workers.

It is becoming increasingly evident that a long-term coexistence of truly cooperative relations and the central-administrative system will be impossible. Either cooperation will develop and expand along with renewal of all aspects of social life, getting rid of extraneous outgrowth, or the administrative system will distort cooperation with all kinds of restrictions and will identify the basic principles and conditions of cooperative activities with the conditions of the present-day state economy.

The development of the economic and social situation shows the growing need for a consistent and comprehensive change in the system of economic relations and above all in relations of ownership by including a number of cooperative principles in these relations. This makes it even more important to foster the further development of new kinds of cooperation and the restitution of cooperative principles of the old form. The main condition for this is the refining of the economic mechanism of the functioning of cooperation, which promotes its specifically cooperative features. The basis of such a mechanism can be formed only by economically and mutually beneficial links between the cooperative sector and the state economy, within which a special role belongs to direct commodity – contractual links, expansion of other decentralized forms of supply of materials such as auctions of all kinds, making available material resources to cooperatives for performing work, providing them with special-purpose credits on a competitive basis, non-dependence of cooperation on departmental structures, the possibility of the operation of cooperatives and their associations without any territorial limitations, etc. The preservation and implementation of cooperative principles squarely raises the question of the character of cooperative unions and their functions and relations with cooperatives. The important thing here is to prevent the subordination of the cooperative unions to the apparat and organs of state management, i.e. turning them into an extension of the management structures and the cooperatives into intra-union economic units by analogy with the existing system of consumer cooperation. One of the factors which stand in the way of such subordination, is the multitude of cooperative unions and the development of their relations with cooperatives on a contractual basis relying on the principles of material, social and other kinds of profitability for cooperatives, which acts as the only criterion of the desirability of participation of a cooperative in one or more unions.

Chapter VIII
The current stage of reform of foreign economic relations

I.P. Faminsky

Director, All Union Research Institute of Foreign Relations; Professor of Economics

1. The Directions of Reform and the Difficulties to be Met

The reform of the foreign economic relations of the USSR, which began in 1986, is now entering a new stage, linked to the acceleration of the transition of the national economy to market principles. Real independence and economic responsibility of enterprises and their status as free producers is the most important part of the concept of transition to a regulated market economy, proposed by the Soviet government. All this is possible only with radical changes in forms of property ownership in our country. A multitude of forms of ownership is an important condition of the transition to a market where demand and supply will influence production structure and the prices of most goods. But competition is as necessary condition for successful transition, in order to succeed in the struggle against a monopoly of producers.

For a long time the Soviet economy was orientated towards developing large-scale production, which led to the concentration of production of many goods in the hands of few enterprises, perhaps as few as twelve. Also, according to some calculations, about 1300 large enterprises are the sole producers of certain products. All this led to undesirable consequences. For example, sometimes monopolistic enterprises imposed outdated equipment upon agricultural users. There are many other examples of this.

Among the measures promoting the transition to the market is the intention to develop a special antimonopolistic programme, which will include stimulation of development of small and medium-size enterprises. It is clear, however, that it will be rather difficult to destroy the monopolistic positions of the large enterprises producing machinery and many chemical products. Here, small enterprises will not be able to withstand the power of the large ones. The real factor acting against this power could be a more active involvement of the Soviet economy in the world market, with a transition to open economy. The current stage of the foreign economic reform must be considered from exactly this point of view.

For foreign economic relations to contribute to the construction of national economy it is necessary to give enterprises more rights here with opportunity to enter external markets as independent producers, the better to understand the world market and to join in competition. The new conditions of marketing will influence the technical standard of products and the quality of their design and manufacture. The state must not merely "allow" enterprises to enter the external market. A special mechanism must be created to "push" them into it. This will force them to become concerned about technical progress in a wider sense.

From the other side, the more active involvement of enterprises in foreign economic activity will help to resolve problems of competition among them on the internal market. The current economic mechanism has led to an unfortunate paradox. For Soviet enterprises it is sometimes cheaper to buy imported equipment rather than home produced machines because equipment bought for hard currency is of better design and quality. This is linked to the fact that monopolies can set the prices of their products on the home market too high. The possibility of entering the external market will provide choice for domestic users. If they are not content with the price and quality of domestic products they would be able to buy them abroad. But to buy abroad, enterprises need currency, hence must increase exports. This would stimulate enterprises to develop competitive products.

Further progress in joint business activity could help substantially in the fight against monopolistic tendencies. At present the contribution of joint ventures to total production is tiny. Also it must be mentioned that industrial enterprises of the Soviet State do not actively want to create joint ventures with foreign participation. It seems that among other reasons there is a fear of losing their monopolistic positions on the internal market. That is why the question of creation of enterprises wholly owned by foreigners, comes up now.

Transition to the market economy requires further reorganization of foreign economic relations. Decentralization and liberalization of foreign economic activity, destruction of current monopolistic structures, and, simultaneously, effective state regulation by economic and legal lever would be the key element in this reorganization. Much has already been done in this direction, but the situation in the national economy dictates further progress in changes towards the open economy.

Prices are one of the most important problems. Internal prices differ radically from world market prices. Internal wholesale prices of fuel and raw materials are about 4% of those in foreign trade. As a result currency earned by enterprises cannot be converted into roubles by the common exchange rate. Therefore, from the beginning of reform in 1987 contract prices have been converted into Soviet roubles using differentiated currency coefficients (DCC). These coefficients were to compensate for the increased expenses linked to production for export and to stimulate export effectiveness, design and quality in accordance with world market requirements. But in many industries especially in those related to fuel and raw materials the old system of subsidies added to the wholesale prices of exported products has been retained. The current system of differentiated currency coefficients has serious deficiencies. First of all there are thousands of such coefficients. They have been repeatedly revised and now they apply to no more than 15% of exported products. The necessity to abolish these DCC and to introduce a common currency rate is evident. The government decree of December 2, 1988 stipulated introduction of 100% subsidy to the exchange rate from January 1, 1990 to replace DCC. But because wholesale price reform was not implemented it was not possible to abolish DCC within the planned period. Thus introduction of a common real currency rate remains one of the most important goals.

This unresolved price problem is one of the main reasons for the introduction of the licensing regulation of foreign operations, and first of all exports. Fuels and raw materials on the internal wholesale market are grossly underpriced and this, with the imperfections of the tax system makes it sensible to place the main proportion of these products in the hands of specialized foreign trade organizations. The large benefits from the export of fuel and raw materials, resulting from the difference between internal wholesale prices and foreign trade prices would go almost entirely to the state budget. If any organizations, including cooperatives, were allowed to export these products they would benefit unfairly and at the expense of state revenues. Further, many export products would escape administrative export regulation.

The price problem also makes the introduction of a new customs tariff, which would really regulate foreign trade, more difficult. The transition to the open economy assumes that this customs tariff would become, as in other countries, the single instrument for regulating foreign economic operations. But under the current price system ordinary duties cannot become a real instrument regulation (for this purpose, duties on some products would be 200–300% or more, which is not accepted in world practice). All these facts show that retention of enormous differences between domestic and external prices is a serious impediment to the integration of our economy into that of the world.

The price problem is one of the key questions to be resolved by the government in the course of the transition to market economy. The government proposes to increase prices on fuel and raw materials, but this would cause a rise in retail prices with serious social disruption. The question of price reform must remain open, but the development of a foreign economic regulation mechanism depends to a great extent on this.

A rather palliative approach to this problem, applicable while substantial difference between internal and foreign trade prices remain, is the introduction of an export–import tax which would correct the difference between low domestic and higher export prices, including those of fuel and raw materials, as well as the difference between relatively low import prices and much higher internal wholesale and retail prices, especially as regards consumer products (for example, audiovisual equipment, personal computers and so on).

The other problem is rouble convertibility, lack of which necessitates a special currency fund system at the level of industrial enterprises. The lack of free convertibility complicates the creation of joint ventures.

The resolution of this difficult problem undoubtedly requires some special steps especially in the internal economy. Without considering them in detail let us outline the main ones. In the first place it is necessary to stabilize money circulation within the country. The lack of balance between the goods supply and the quantity of money, i.e. the internal market deficit, really means there is no convertibility of rouble even on the internal market.

Secondly, it is also necessary to resolve the problem of making domestic prices approach world market ones. The introduction of convertibility under current prices would lead to enormous speculation.

Thirdly, it is necessary to introduce a well founded rouble exchange rate. An understated exchange rate would reduce the foreign currency holdings needed for internal requirements, but at the same time the purchase of many products would only be profitable for foreigners, which would result in a large outflow of our national resources. The exchange rate must be reasonable.

Specialists would calculate the exchange rate not on the basis of average correlation of prices (the current rate, taking into consideration the very low wholesale prices of fuel and raw materials which are our basic exports, corresponds to an average correlation of internal wholesale and world market prices), but on the basis of the so-called limit. It must guarantee that the bulk of exported goods are profitable to internal producers. From this point of view it must be 40% of the current rate, if calculated on the basis of the present DCC for machinery. It was stated in the government report to the Supreme Soviet that the change of rouble rate, to make it correspond truly to other currencies, must be done this year.

The change of exchange rate is an important step to convertibility of the rouble. In the future in unison with creation of market economy and transition to rouble convertibility this rate must react quickly to internal price dynamics and to the world currency situation.

Fourthly, for ensuring convertibility, the current poor state of our balance of payments must be overcome, our relatively large foreign debt must be reduced and exports must be diversified for currency earnings not to depend, as now, exclusively on one product alone (fuel and petroleum products).

The conditions for transition to full rouble convertibility will surely not be created at one step. This process will take time and will depend substantially on the speed with which the market economy develops in our country. However, it seems that certain steps towards convertibility and a currency market must be taken now. Currency auctions conducted from the end of 1989 can play a certain role. The high prices of foreign currency at these auctions (the price reached 35 valutny roubles for 1 dollar, i.e. more than 20 roubles to the dollar) can be explained by the fact that only a few enterprises offered hard currency for sale, and at the same time, the demand is very high. The significance of auctions as one of the channels for currency earning and at the same time as a mechanism for creating a currency market in the country can be increased by offering some centralized currency funds at these auctions. Joint ventures must be permitted to take part in these auctions, which will help to resolve the problem of repaying their currency. In the future it would be possible to go from auctions conducted periodically to a constantly operating currency bourse or a regular trade in hard currency at the market rate, accessible not only to enterprises but also to individuals, which will signify the creation of a currency market and the introduction of internal convertibility.

The necessity of accelerating the change to use of market instruments for currency regulation is linked to the radical reform of economic relations of the USSR with countries of the CMEA. From January 1, 1991 our trade with these countries will be done on the basis of hard currency and present world market prices. For a long time, the relations among countries of the CMEA were founded on the average world market prices during a certain period. At the beginning, the prices were fixed for the next five year period on the basis of the average prices during the five preceding years. From 1975 prices were set every year, on the basis of average prices during the five preceding years. The same mechanism now, in 1990, has helped East European countries to adjust gradually to new high oil prices during periods of energy uncertainty, and the USSR to adjust to lower oil prices after 1985. However, as a result of this mechanism, goods imported from East European countries were overpriced compared to real world market prices, especially as regards the machines and equipment that make up almost half of Soviet imports from these countries.

Accordingly to estimates of Soviet and foreign economists the annual losses suffered by the USSR as a result of this, are 1012 billion roubles. The USSR has a deficit in trade with the East European countries (2,8 billion roubles in 1989) because of the slump of oil and petroleum products prices. With a transition to real world market prices the correlation will change naturally in favor of the Soviet Union. But it would be possible to use world market prices in trade with East European countries only after Soviet customers with currency have become able to choose home produced or imported goods on a competitive base, i.e. once the relations with the East European countries are of a real market nature.

There are still major problems with the country's deficit in consumer goods and in the means of producing them, and the related contemporary problems of reform of the domestic economic mechanism, the system of reserved state orders, which sometimes account for a considerable volume of production, and the absence of a retailing trade, complicate further reform of foreign economic relations.

If you look at what real possibilities there are for an enterprise to export more finished goods, you will find out that very often the main obstacle is that the enterprise is not yet in command of its own production. Not only are there state orders, but also enterprises must conclude treaties on selling their products to home consumers, and these treaties are often concluded under pressure from the top. Under such a situation they have no production for the world market.

This kind of situation arises, to our regret, from a failure in the country's economy. Many proclamations have been made advocating the rapid development of foreign economic relations and a transition to open economy, but implementation of these principles does not follow. Moreover, sometimes measures are introduced that lead directly to opposite results. Among them, for example, are restrictions on the exports of industrial goods that were an element in the measures on stabilization of the Soviet economy, adopted by the USSR Supreme Soviet in November 1989.

Still persistent is self-centered thinking on the part of the public, that one can observe in the newspapers. From time to time this or that newspaper begins to campaign against exports of commodities that are lacking on the domestic market. But now there are deficits of both industrial and consumer goods. If you take the road of reducing exports, further isolation of the Soviet economy from the world economy may result, retarding it further and doing nothing to remedy technical backwardness and low quality of production.

2. The Increasing Role of Business in Foreign Economic Relations

Reconstruction of the organization and management of foreign economic relations in the USSR, through application of many organizational and economic measures has changed radically the position of enterprises, unions and organizations (the mainstays of the country's economy) in foreign economic activities.

According to the Decree of the USSR Council of Ministers (December 2, 1988) all state, cooperative and public enterprises and organizations now have the right of direct operation of exports and imports. By June 1990 more than sixteen thousand state enterprises, cooperatives and other organizations have been registered as participants in foreign economic relations.

The enterprises and production unions which export and import on sufficiently large scales, create their own foreign trade firms (FTF). In some cases such firms are represented on, and preserve tight business relations with, sectoral foreign economic organizations. Some FTFs operate not only the foreign economic relations of the parent enterprise, but also its subcontractors, as takes place, for example, in the commercial activities of the "AutoLada" firm.

Firms within a number of intersectoral scientific and technological complexes (ISTC), including the ISTCs "Microsurgery of the eye", "Mechanobr", "Biogen", etc. have successfully begun foreign economic activities. Successful foreign economic activities of ISTCs depend directly on the fruitfulness of their main scientific, technological and economic activities. In this respect the foreign economic activities of the firm "Mikof" of the ISTC "Microsurgery of the eye", that has begun to participate in investment activities both in the USSR and abroad are characteristic. Development of foreign economic activities is made easier in practice by creating intersectoral industrial trusts ("Kvantemp", "Technochim", etc.) to include intersectoral foreign companies.

Under the new conditions Soviet enterprises will take an ever more active part in creating joint ventures on the territory of the USSR and shareholding societies abroad, and this opens for them new possibilities for integration into the world economy, such as creation of their own production network and better access to the markets held by their foreign partners in such joint ventures.

It is quite natural that provision of the right of direct access to the export market does not mean that every enterprise will create its own foreign trade structure and operate independently in the foreign market. This right enables an enterprise to choose the best way to sell and purchase commodities on the external market, and to use for that purpose those foreign trade organization will provide it with the best conditions.

For the role of enterprises in the foreign economic sphere to increase it is important not only to present them the right of access to the external market, but also to create a system of export promotion, and stable conditions for their general foreign business.

While we lack rouble convertibility a major priority of exporters is the accumulation of currency funds. Under conditions of a deficit interior market the strong interest that enterprise collectives take in exports arises partly from the possibility of access to necessary import commodities. In the existing system, foreign currency funds of enterprises usually come from payments made according to fixed rates from the currency profit by selling finished products and services. Currency payment funds, excluding the part that goes to ministries and departments and also the 5% for local authorities, cannot be taken by directing authorities and is used by enterprises themselves. At the present time many limitations of enterprises' currency means are being eliminated. Up to 25% of the enterprises' means in convertible currencies is permitted to be used for purchasing consumer goods for resale to workers at these enterprises.

However the existing system whereby currency funds of enterprises are formed, based on sectoral rates, is not free from faults and is supposed to have been improved by the beginning of 1991. For example, at present the system lacks the selectivity needed. This is because exports of highly processed commodities are being stimulated. To make the best use of limited currency funds such techniques as barter operations, border and offshore trade, or exporting in cooperation or in joint ventures with enterprises, with higher sectoral rates of currency payments, can all be used.

The mining industry formed their currency funds by exporting production surplus to planned targets, but for some types of commodity there were absolutely no payments to currency funds. There is also no currency payment in favor of enterprises which export products according to government treaties through state credits of a part of gratuitous aid programmes.

To discuss the formation of currency funds for enterprises we must presume firstly, creation of currency funds for enterprises throughout the economy; secondly, to move from the sectoral to the commodity principle of calculating currency payments rates, meaning leaving most of the currency to enterprises which add value to the commodity in the process of exporting; thirdly, establishing a single principle of payment of a definite part of the currency profit to centralized republican and local currency funds.

Currency payments are being quantified with reference to the country's needs in centralized currency expenditures (for payment of external debt) and also to the division of functions between the Union and Republics in the foreign economic sphere.

It is presumed, too, that the existing rigid borders of currency self-financing of enterprises will be relaxed. It is necessary for that purpose to distribute some centralized currency means on a credit basis through the bank system, and also to form a currency market, which will mainly operate as currency auctions.

For foreign economic activities to have a positive influence on general results of businesses the renouncing of differential currency rates and the introduction of a real rouble rate will be of vital importance. But the considerable differences in profitability of exports and imports of different commodity groups in different industries mean it is impossible fully to abandon subsidizing loss-making export–import operations. But gradual reduction of state subsidies and their future elimination are possible provided this is done in a context of rigid of measures aimed at reduction of costs, better quality and greater competitiveness of exports.

A system of export credits has a considerable importance for the expansion of export activities of enterprises. Such systems have been adopted widely in advanced countries. It is also necessary to extend hedging of export and import operations, as well as export credits. These processes have only begun to develop, and must be accelerated.

Training of qualified personnel will be important for an increase of foreign economic activities. During the last 34 years a lot has been done in this area. The preparation of students specializing in foreign trade is broader, in-service training has been organized, business schools have been created, visits of specialists abroad have been extended. But it is the low quality of personnel that remains the weakest point in the foreign economic activities of enterprises.

Advertising has a part to play in the development of foreign trade, but development is slow. Advertising agencies have been unable to cope with the demands.

A big problem for participants in foreign trade is absence of information on market prices etc. At present this is one of the major weaknesses of our foreign trade system. In this connection the creation of an Association of business cooperation for the exchange of foreign economic information would be of considerable importance. This association could accumulate information and transmit it to export–import businesses on a commercial basis. Such an association could be organized on a regional footing. Cooperatives and joint ventures can also play a more active role.

3. A New Role for Trade Intermediaries

In spite of the growing role of businesses in foreign trade the main volume of Soviet exports and imports is through foreign intermediary organizations fulfilling export and import operations on the basis of contracts and commission. During the years of reform these foreign economic organizations have changed and improved considerably. In the process of transition to market economy we'll have to overcome the existing elements of monopoly here.

It is known that till 1986 all export and import operations had been monopolized by the All Union foreign trade organizations, being part of the system of the Ministry of Foreign Trade (MFT) and USSR State Committee on Foreign Economic Relations (SCFER). Every such organization had its own allocation of export and import commodities and services.

In the process of reform, the network of foreign trade organizations has been extended.

First, the structure and composition of the biggest trade intermediaries (the All Union foreign unions) have changed. Their major part (25 unions) is still subordinate to the Ministry for Foreign Economic Relations. They export and import on behalf of the state, organize economic and technical aid to foreign countries, organize jointly with other ministries and departments construction projects abroad and those on Soviet territory with participation of foreign organizations and firms. They transact nearly 60% of Soviet foreign trade.

Second, foreign economic organizations are being created in practically all industrial ministries and departments. The MFT and SCFER of the USSR have dedicated to creation of these unions, whole unions and specialized firms. New unions can use the Ministry apparatus for organization of foreign deliveries.

Those organizations have to connect foreign trade more tightly with industry, to permit more centralized development of the export potential of businesses.

Third, further decentralization and democratization of organization and management have led to independent foreign economic organizations attached to the Councils of Ministers of Union republics.

Their territorial specializations have simplified considerably the access to foreign markets of enterprises, especially those of republican and local affiliation, and the involvement of additional export resources in foreign trade.

New foreign economic organizations have been emerging. A dozen newly created economic associations engage directly in export and import. These associations are becoming more and more numerous. They are being established by groups of enterprises of the same kind and in the same region and city. Such associations are especially important as they can mobilize the export potential of medium-size and small enterprises. Major industry or republican companies consider servicing medium-size and small enterprises unprofitable because small deals take no less time to prepare and carry out and no less effort by the experts than large contracts do.

Associations promoting business cooperation with foreign countries occupy a special position within the range of organizations acting as agents. They are designed to assist Soviet enterprises to develop entrepreneur activities as well as new forms of foreign economic cooperation. Associations of this kind are oriented to certain countries, regions and industries and unite corresponding state enterprises, cooperatives and other Soviet organizations of all kinds and geographical locations.

Such associations are established with India, Italy, Federal Republic of Germany, the SFRU and China, with Pacific and Latin American regions. There is also an association formed to develop the Kola Peninsula's natural resources. Branch associations have been established in the agro-industrial sector, in construction and in the production of construction materials.

These associations render practical service to their members as well as to other Soviet and foreign interests helping them to find partners for joint ventures (JVs), to prepare commercial negotiations, to conduct market research, to conclude contracts and to provide them with various consultancy services. The associations are in a position to represent their members abroad on a commercial basis.

The transition to the market economy and the emergence of different forms of property cannot but affect foreign economic relations. Already proposals have been made for joint-stock companies based on the foreign trade organizations (FTOs) and subordinate to the USSR Ministry of Foreign Economic Relations. They may also be based on the branch foreign economic organizations. There are also proposals on ways significantly to enlarge the subject and the scope of their activities. Establishment of foreign trade joint-stock companies means that the above mentioned FTOs will leave the control of the Ministry of Foreign Economic Relations and of the branch ministries to become autonomous trading agents. Emergence of foreign trade companies, foreign economic associations, and republic-based organizations will result in the FTOs losing their allround monopoly in certain sectors of foreign trade. Their transformation into trading companies with a much larger range of specializations will help to put an end to their monopolies in corresponding spheres of foreign trade.

A problem has emerged in connection with FTOs of specialized profile becoming re-established as trading companies. Some such companies incorporate banking, insurance and transport enterprises along with the foreign trade organization as well as enterprises dealing with wholesale and retail trade. Scientific research organizations, information organizations and certain industrial enterprises may also be included. As a rule, companies of such a kind are established as joint-stock companies incorporating JVs, industrial cooperatives and other organizations located not only in the USSR but also abroad.

During reform it has become possible to improve the contracts between the FTOs and export–import enterprises. At present such relations are developing on a contractual basis, with contracts of supply, agency and commission being concluded in particular. Here enterprises can set terms for goods and services at their own discretion.

In connection with this, measures have been taken to develop the FTOs' self-finance system. Having once been agents, the FTOs used to be financed from the state budget and did not have significant fixed and current assets (or foreign currency either). The difference between export prices and internal market prices was referred to the state budget. Thus the FTOs had no means to finance their own economic and commercial activities and were not interested in the development of foreign economic activities.

In 1990 the FTOs have really started to implement the principles of self-finance which means that their costs are to be covered by their own profits. Nowadays commission rates are charged by contracting parties and corresponding amounts are to be paid by enterprises. This mechanism must be further developed.

The deductions made for export and import services of all kinds still remain equal for each organization in the foreign economic relations area. This does not stimulate them to improve the geographic structure of exports and imports, or to choose the most effective terms of sale and schedules of payment. Moreover, the general method stipulates that the amount to be charged as a commission must be calculated considering equated interest which depends on the total volume of goods and services having been realized within a certain period of time. This factor makes the FTOs differentiate deals according to their potential profit. In particular, the FTOs consider it unprofitable to engage in dealings of only small value. Along with this, another tendency has been revealed; monopolistic FTOs claim excessively high commissions for deals though such claims are groundless. Current forms and methods of payment to the personnel of FTOs do not stimulate their interest in trade growth.

Due to the considerable increase of numbers of foreign trade organizations and firms, and to changes in their structure, the development of more elaborate forms of their self-finance and their financial incentives have become acute. Bearing in mind first of all the principle of self-finance must be implemented when compensating for the costs incurred by the FTOs in effecting individual deals or groups of deals (calculations are to be made both in roubles and in foreign currency). These amounts are to be referred mainly to the enterprises' accounts. Consideration must be made of the volume and quality of the service rendered (namely, export of goods) as well as of its effectiveness and value. Relations between all the FTOs and the enterprises supplying goods for export and placing orders for imports should be developed on a true contractual basis while the types of contract which are practiced at present should be improved by considering real terms of agency or supply contracts. Establishment of joint-stock companies on the basis of the FTOs will naturally speed up this process.

4. Conditions for Foreign Capital Investments

Foreign capital investments are to play an important role in the reform process helping to solve both the problem of the USSR internal market and that of diversifying the range of export goods. By May 1990 about 1.800 JVs have been registered while the total amount of the corresponding capital investments exceeds 1,6 billion roubles. Still we are not yet satisfied with the development of this process. Few JVs have been established in the industrial sector, especially in machine building. Involvement of advanced foreign technology has not yet achieved any significant level.

There is a whole range of problems impeding expansion of foreign capital investments, e.g. nonconvertibility of the rouble, lack of wholesale trade as far as means of production are concerned. Problems of industrial, business and social infrastructure development remain to be solved, especially in remote districts where it also impedes establishment of JVs. There are problems relating activities of JVs which must be settled by legislative measures. A law on foreign investments is being prepared to eliminate the latter group of problems.

The bill stipulates development of new forms of foreign capital investments. Foreigners will be allowed to invest not only in those enterprises located in USSR territory in which Soviet legal entities and persons participate (i.e. in joint ventures) but also in enterprises owned exclusively by foreign investors.

Agreements on land, building and equipment lease will also be allowed as well as agreements on concessions.

Foreign investors may engage in any sphere of economic activity except those to be included in a special list. The radical program of transition to the regulated market provides perfect opportunities to develop the capital market. A major programme aimed to transform state enterprises into joint-stock companies and to develop a securities market is being outlined in the USSR. It is also planned to give firms the right to acquire securities.

The emergence of joint-stock companies in the USSR permits foreign capital to be invested in diverse forms. Foreign capital will be allowed to participate in joint-stock companies, in limited liability companies, in other economic associations and in partnerships.

The USSR Supreme Council has approved a property law making a legal basis for the establishment of companies with a charter fund to be formed exclusively from foreign capital. Generally, these enterprises are to be operated under the taxation system practiced for the JVs.

The basic law on leasehold of the USSR and union republics having been approved by the USSR Supreme Council, the lease of property to foreign investors and to the JVs is permitted while the lease of land and any other natural resources is to be regulated by decrees of the Council of Ministers and by decrees of the autonomous republics owning territories where property is let on lease.

Concessions are to be granted for no longer than 50 years on the basis of agreement on the following: schedule of payments to be made as a lump sum and annually; system of financing construction of dwelling houses and of social infrastructure; participation of leasehold proprietors in concession management bodies; system of compensation for ecological damage resulted from utilization resources by concessionaires.

Problems of joint venture zones (JVZs) are being actively discussed in the USSR as these zones might attract foreign capital to develop certain regions. JVZs are to be subject to Soviet legislation, and organs of Soviet power; Soviet enterprises and organizations will operate on their territories. At the same time there will be favorable conditions for foreign capital regarding taxes, leasehold norms and duties. It is essential that the necessary industrial and social infrastructure should be developed. JVZs must not become exporting enclaves of the kind found in developing countries but they must also deliver considerable volumes of their products to the internal market of the USSR. JVZs might be organized first of all in the Far East, for instance in Nakhodka. The most promising zones seem to be those where foreign capital will be engaged in assimilation and practical implementation of Soviet scientific and technological achievements (in the West a zone of this type has been called a technopolis). A similar project is planned for Vyborg. Proposals have been put forward concerning establishment of JVZs in Kaliningrad, in the Chita region, and in Sakhalin. The RSFSR Supreme Council has approved the idea of such zones. Market relations will develop in JVZs in without restriction, thus they will provide important links between the USSR and the world economy.

The problem of capital exports from the USSR has been emerging as well as the problem of establishment of JVs abroad and the problem of foreign stock acquisitions, etc. The USSR government has made basic decisions in connection with these issues. But as long as Soviet enterprises do not possess adequate foreign currency it is hardly reasonable to expect any major steps to made in that direction though, since the prospects are generally good, cooperation between the USSR and foreign countries in the industrial sphere may be significantly stimulated. Soviet enterprises have already made deals to acquire foreign stock.

Revolutionary changes of the economic model have naturally imposed the task of promoting a clear understanding of the essence of the market, and of its positive and negative aspects. A change of public opinion towards all aspects of the market as well as towards its elements (money market, securities market, investments market) is required. Particular issues of credit and price reforms, and of a new rouble exchange establishment must be solved as they are of significant importance. The majority of foreign firms are still examining the USSR as a potential joint venture partner and as prospective area for capital investment. Meanwhile the USSR with its material and human resources may be considered as an important element of the world market, moreover, its present foreign economic strategy is aimed at further integration with the economic structures of the world.

Chapter IX

Problems of radical reform of pricing system in the USSR

Yu.V. Borozdin

Professor of Economic Sciences, Central Economic-Mathematical Institute of the USSR Academy of Sciences

The current economic reform in the USSR presupposes renunciation in principle of the administrative command system of economic management and a transition to planned-market management using economic methods and tools of trade and fiscal relations, such as prices, money, finances, profits, credits, etc. Yet many still take a rather formal view of this widely held and seemingly incontrovertible belief assuming that to accomplish a radical economic reform it would be enough just to modify the role and the importance of commodity-money categories under socialism and to bring the system of prices, finances, credits and people's incomes into conformity with present-day conditions of economic development. This would leave the administrative command system of management basically intact with only some minor adjustments being made "in the spirit of our times". However it should be borne in mind that if the administrative-command system of the Soviet economy management is not dismantled fully and completely, no economic reform whatsoever will come to fruition in the country, let alone a radical reform.

The present pricing system represents an element of the administrative command mechanism of economic management and therefore incorporates all its negative features. Objectively speaking, it is production cost-oriented in character, promotes economic relations based on barter deals, is strictly centralized and bureaucratized, and, for all practical purposes, absolutely ignores market demands. The pricing mechanism is in large measure administratively controlled, stringently regulated and therefore extremely inflexible and rigid. By its very economic nature, such a system of prices and pricing cannot help to enhance the effectiveness of public production, stimulate scientific and technological progress, promote the rational use of all production resources, balance demand and supply, or help in the choice and implementation of optimum economic options.

There are two fundamentally different approaches to a comprehensive overhaul of the system of prices and pricing in the Soviet Union. The first one is predicated on the existing practice of one-step across-the board reviews of prices made by administrative pricing bodies which set prices in a centralized fashion depending on the actual production cost of goods. The second approach is based on a general concept of economic reform providing for a transition to market relations, and consequently, it rejects in principle any administrative price-setting and calls for the establishment of a system of equilibrium prices bringing demand and supply into balance, i.e. market prices. The road to those prices will be a very difficult one, but it has to be travelled.

At the same time, whatever the general approach to the reform of the pricing system in the USSR, it is necessary to correct as soon as possible unreasonable proportions between prices of raw materials and fuel and those of manufactured goods. The imbalance here is obvious from all points of view. Unrenewable natural resources should be assessed not only according to the actual social cost of their extraction, but taking into account the most rational and frugal way of their use, as well as the overall strategy of structural policies designed to reduce the relative share of the raw material sector and increase the share of manufactured products in the gross national product. Prices of raw materials should serve to constrain their consumption by industries, while prices of manufactured goods and finished products should stimulate enhanced production and consumption thereof. In this respect, world price proportions can be taken as fully acceptable points of reference for us. But only as points of reference, in contrast to the view that an overall pricing reform needs to be carried out in such a way as to make the level of our domestic prices coincide with existing world prices.

This view is contestable not just because the world prices of the overwhelming majority of specific types of product is a highly vague notion, but because world market prices of major commodity categories reflect the actual state of the world economy today. Our national economy differs in many parameters from the world economy which is shaped primarily by the economies of leading capitalist countries. An administrative change-over to world prices is conceivable and even feasible, but no administrative means will enable us to bring our economy up to the world level. After all, this would involve the corresponding restructuring, raising the quality of produce to meet world standards, improving the structure and mix of supplied goods, increasing per capita incomes, etc. It is only on this basis that we would be able, over time, to make the rouble convertible. Thus the task of harmonizing domestic prices with world prices cannot be addressed separately from the overall problem of creating conditions within the national economy that would help to bring it as close as possible to the world economy. This means that, in the first place, an economic mechanism should be set up to start movement in that direction. And that mechanism cannot be based on the administrative-command principle even in its non-essential parts.

A radical economic reform has to bring about fundamental changes in all principal components of the management system, i.e. in planning, material and equipment supply, pricing, financing and crediting, and also in remuneration for work. What is at issue here is changing the economy over from administrative-command methods of management to economic ones that would be fully geared to socialist market requirements and at the same time would not preclude the state from influencing major socio-economic processes.

It is common knowledge that all the above-listed elements of the management system in an administrative-command economy are closely interrelated and match each other, so they can be reformed concurrently and all together. It is impossible, for example, to maintain directive planning in physical terms while switching over to free-market prices. By the same token, quota-based or rationed distribution of products is incompatible with the planning and financing independence of individual production entities. Plainly speaking, if we are to effect a radical economic reform, we need a clear concept of what the eventual economic mechanism should look like and what action must be taken to build that mechanism. As yet there are no grounds to speak of full clarity on these matters, although until recently a majority of Soviet economists seemed to have no doubts about the basic principles and approaches to the implementation of a radical economic reform. Of course, while differences existed on the content and objectives of such reform, no one challenged the need for a profound and coherent restructuring of prices of all kinds – wholesale prices of industrial goods, agricultural and consumer prices, as well as transport and service tariffs. And this is fully understandable because the price system has a vital function in the entire economic mechanism, and a new economic mechanism cannot be "set in motion" under the old price system which has evolved and continues to operate on the principle of cost-oriented and administratively regulated pricing – a principle that is absolutely typical of a command economy. So let us formulate some basic ideas. Firstly, radical economic reform is not feasible without a reform of prices and pricing. Any attempt to defer a price system reform, whatever the excuse, would be tantamount to the abandonment or suspension of the entire economic reform. Secondly, the revamping of the system of prices and tariffs should involve them all and be carried out in a concerted and consistent manner. Thirdly, the overriding objective of an across-the-board price reform is to dismantle the mechanism of administrative price-setting by government authorities and to establish an effective mechanism for price formation in the socialist market in accordance with the laws of its operation.

As you can see, a price reform calls, on the one hand, for changing prices and tariffs and bringing them more in line with the socially indispensable production costs, the usefulness of goods and services and the existing correlation between demand and supply, and on the other hand, for reforming the pricing mechanism itself through the renunciation of the practice of setting prices "behind closed doors" on the basis of actual production costs, and through a consistent transition of free price formation in a socialist market.

Formally, many agree on this price reform concept, but they differ substantially on practical ways to materialize it. Of special interest in this context is the approach taken by the USSR State Committee for Prices as the agency in general charge of the price reform.

The thrust of that approach is to make prices contingent on actual production costs without taking into account their social usefulness, and arithmetically to recalculate prices after cost-increments along the whole chain from initial raw materials to end products while fully disregarding real consumer demands, fixing prices for millions of goods and services through rigidly centralized administrative-bureaucratic methods, and barring the consumer from the price formation process.

Leaving in place the administrative pricing mechanism, even if decentralized by the so-called delegation of price-setting authority to ministries, agencies and individual regions in the country, would make it impossible to conduct a radical economic reform. Why so?

First, prices established by orders from on high rule out truly free economic action both within enterprises and in economic relations among enterprises since such price formation pattern can, in effect, only be based on production costs. This means that prices would not take into account fluctuations and would only be a reflection of the profit rate fixed by directive at a level sufficient for compulsory payments to the budget and for maintaining economic incentive funds.

Second, administratively regulated prices put enterprises firmly in the position of an applicant vis-a-vis the price-setting bodies that request prices meet manufacturers' vital needs. However, if administratively fixed prices fail to perform that function, the enterprises do not stand to lose either as they are given subsidies anyway.

Third, administratively set prices presuppose that production planning and distribution of products occur outside enterprises and without their participation thus precluding their genuine independence. Therefore no need arises for switching from the plan directive to the plan-order.

Fourth, administratively regulated prices render superfluous trade in finished products, for if prices are fixed by authorities, then products, in turn, should be distributed according to quotas and rations. This virtually rules out economic relations between manufacturers and consumers.

Fifth, administrative regulation of prices can be carried out separately in every price category, including wholesale, purchasing and retail prices, and in isolation from the plan since prices are not intended to serve as major tools of the socialist market and to maintain a balance between demand and supply. This is why any local readjustment of prices can be made without changing all price proportions just through a simple arithmetical calculation of upward and downward shifts in prices.

So the present-day content and techniques of pricing, the centralized adoption of hundreds of price lists and the bureaucratization of the pricing process proper are incompatible with the requirements of radical economic reform. What is needed here is a major transformation involving not only a review of prices and tariffs throughout the economy, but a fundamental change in the pricing mechanism. This is all the more important to note since the growth of inflation with every passing year exacerbates commodity-cost imbalances in the national economy. Prices of products follow, albeit with a delay, the amount of money in circulation and, in turn, stimulate a new spiral thereof. Since 1967, wholesale and purchasing prices as well as costing estimates and price rates in construction have been increasing consistently. To be noted among particularly important steps in the field of pricing have been the review of wholesale industrial prices in 1982, of purchasing price of agricultural produce in 1982 and 1983, and of construction costing estimates and price rates in 1984.

It has to be pointed out that all price reviews in the last two decades have followed a pattern strictly based on production costs – every markup in the cost price in extractive industries caused increases in raw material prices which triggered a technological chain reaction, resulting in greater production costs and prices in all other sectors of the economy. The cost-based price model consisting of the cost price plus the planned profit rate is an inherent feature of a cost-oriented economic mechanism relying on administrative-command methods of management, gross output targets and progressively planned output growth over the achieved level. Typical of this mechanism is the equating of costs with results, and therefore any growth of costs in the economy and of product price estimates not related to the consumer product's qualities was not regarded as a negative development but rather as an economic boon. It was the production expenses criterion that also underlay the way in which the cost of surplus product was taken into account in the consumer price, as the profit reflected in specific product prices was calculated in percentage points of the prime cost or its modifications.

In recent years the deficiencies of cost-oriented pricing and of the cost-oriented economic mechanism as a whole have drawn particularly strong criticism in the USSR. There is no need to rehash it here, yet I would like to say one thing – letting prices hinge on production costs is a road to nowhere following which we would not be able to solve our highly complex problems. Rather, by keeping the cost-based pricing system in place, we would only aggravate the crisis of our economy.

Prices in a real market economy perform their genuine economic function only if they serve as a vital tool for maintaining a balance between demand and supply. By ignoring this principle it is impossible to break the vicious cost-price circle.

The situation that is taking shape now is, in effect, the same as the one that existed in 1967 and in 1982 – increased production costs cause the emergence of loss-making and low-profit production of fuel and raw materials resulting in the growth of prices of these commodities, and this growth goes on along the chain of production links through all other sectors and returns to the fuel and raw material sector, albeit at a new, higher level, thus making a new price reform necessary. No wonder therefore that the very same problems that were addressed during the two earlier most significant price reviews in 1967 and 1982 have now reappeared again. They include the unprofitability of the coal industry and the extremely low profit rates in some other mining industries, obvious imbalances in prices of products of extractive and manufacturing industries, unjustifiable disproportion in profit rates, responsiveness of prices to requirements of scientific and technological progress, enterprises' lack of interest in economizing and saving resources and, by contrast, their drive to send prices up by all means, fair and unfair alike, etc.

As we have noted earlier, these cannot be solved if the old content and methods of price-setting practice are preserved.

For more than 20 years pricing policies for consumer goods and services have been based on the principle of maintaining price stability and whittling prices down in an economically reasonable manner as corresponding commodity stockpiles are created and production costs are cut. But due to developing inflation it has not proved possible to follow this principle consistently. While the level of prices of some staple foods was kept unchanged, there was an active hidden growth of prices underway as a result of reducing the choice of inexpensive manufactured goods, increasing the average price of each purchase, and allowing faster price growth than the quality justified. The scarcity of consumer goods in the market was on the increase, which made itself manifest in the exacerbation of the overall and structural imbalance between consumer demands and the supply of goods and services.

At the same time, the political objective pursued during all these years was to maintain stable consumer prices which, given the active policy of raising the population's nominal money incomes, was supposed to bring about an adequate increase in real incomes and a higher living standard. However in practice this was not to happen because the hidden price growth partly dampened the effect of raised nominal incomes, and the widening gap between supply and demand spawned a series of other negative phenomena, such as a slump in the rouble's purchasing power, revitalization of the second (shadow) economy, mushrooming speculation, etc.

Besides, there was an increasing deformation of the entire pricing system – the level of retail prices of many food products proved lower than the level of wholesale and purchasing prices, which lead to annual increases in government subsidies from the USSR state budget to offset the difference between retail prices and actual production costs – subsidies that totalled 103 billion roubles in 1989 – an increase of 12.6 percent over 1988. Out of that amount, 87.9 billion were food subsidies.

All this was a consequence of the selective approach to the interrelated price system and of disregard for the logic of real economic development.

The task of carrying out a sweeping price reform in the Soviet economy presupposes not only a radical and coherent restructuring of prices of all kinds – wholesale, purchasing and retail – but a change in the price-setting mechanism itself. What is reform's goal? The answer to this key question will in great measure determine the content and methods of the prices and tariffs review in all sectors of the economy and the very concept of the new role that prices will have to play in a market economy.

Until now the declared aim of all general as well as partial price reviews in the Soviet Union has been to reflect in prices as fully as possible the required labor input. Because the Soviet price-setting practice meant prices are in effect cost-based, the purpose of all price reforms attempted so far has boiled down to the fashioning of new prices following the current production costs existing at the moment of the reform's introduction. Thus the objective was transformed into a price-setting principle that could be easily implemented in any isolated price area, such as wholesale prices of manufactured goods, purchasing prices of agricultural produce, transport and communication tariffs, or construction price estimates and costs. It was also no accident that the sequence of all new preparations of price lists has always strictly followed a pattern starting with prices in basic, fuel and raw material industries and moving on to prices of intermediate products of the manufacturing industries and further on to the wholesale prices of finished goods. Under an economic mechanism where barter relations prevail and prices serve, in fact, only as means for measuring current production expenses, such pricing techniques are quite natural and logical. After all, prices in this case are only needed for overall cost gauging purposes, and have no real effect on the production process, assessment of the effectiveness of business activities for the cost–result ratio.

Now there needs to be a fundamental change in the situation. An economic mechanism of a new type based on economic methods and market relations requires adequate economic tools to replace administrative-command levers of management. Besides prices this includes a full-blooded money circulation system precluding any major difference between cash money circulation and clearing money circulation, fixed payment rates for resources and profit distribution norms, wear-and-tear standards and interest rate norms, tax rates, the size of penalty sanctions and bonus payments for economic performance, etc. This system of economic regulators is called upon to harmonize the interests of the society as a whole and those of the collective and the individual worker with a view to attain best results at a minimum cost. It will be the backbone of an economic system that will make it possible to translate into practice the principle of the basic production unit's economic independence which finds its expression above all in the fact that the unit draws up and implements its own short- and long-term production plans. This is the only way to make a transition from the administrative plan-directive system to the economic plan-order system. Under this system, prices can no longer be set administratively, by authorities; they can only be formed in direct interaction between producers and consumers, that is, in effect, contract prices striking a balance between demand and supply. If prices of the bulk of the range of products and services available are set exogenously rather than in direct or by-proxy contracts between producers and consumers, there can be no hope for the dismantling of the administrative-command system of economy. But before a transition is made to such a self-adjusting economic system capable of certain self-regulation using the entire set of economic tools, a number of radical measures in economic management need to be taken.

These measures will, no doubt, include an across-the-board reform of prices and pricing. For that matter, in contrast to all earlier improvements in individual price categories, the now-planned reform, we think, should follow a fundamentally different scheme starting with end product prices (consumer prices), proceeding to manufactured goods prices and coming finally to prices of basic raw materials. This is exactly what lies at the heart of a reproductive approach to pricing and to the organization of the entire system of economic management in conditions of real commodity-monetary relations. However, these relations serve as a typical evidence of the existence of a socialist market with its inherent features, one of which is the quest for a state of equilibrium.

An unbalanced market where demand is not met by supply bears witness to disproportions in the national economy's development and, consequently, to the decreasing efficiency of the economic system as a whole. This has been exactly the case with our economy for many years. Short supply is typical not only of the consumer goods market, but of the commodities, machinery and raw materials markets as well, although the latter, strictly speaking, are not markets at all since they are dominated by a system of fund distribution and rationing. Incidentally, this explains why, despite the enormous production output in many mining and manufacturing industries which substantially exceeds the volume of similar production in other economically developed countries, a chronic shortage of goods is constantly felt. The USSR leads economically developed nations in terms of extraction and production of coal, oil, gas, steel, rolled metals, cement, mineral fertilizers, machine tools, tractors, combine harvesters, etc., and yet these items remain in short supply. And the reason is not merely a physical imbalance between resources and demand; such an imbalance theoretically can be redressed through in planning, but in real economic life it cannot be removed until physical production targets in the system of "production-distribution-consumption" are replaced with economic relations between sellers and buyers using the money they earn and given prices that promote a balance between demand and supply.

If resources are allocated to enterprises by authorities on the basis of quotas instead of being acquired by enterprises in the market using their own funds, they will always be in short supply because the volume of resources requested by enterprises today is not limited in real terms by the enterprises' own financial capabilities. Rather, the opposite is the case: once resource quotas have been allocated, governmental subsidies to cover them will be made available to enterprises anyway. And so long as there are no real and equivalent monetary relationship between exchange participants whereby they pay their own money for products they need, shortages will persist. In the same vein, one prerequisite for a real and economically sound management should be considered to be economically substantiated prices helping to balance demand and supply in a socialist market.

So the purpose of a radical price reform in creating an economic mechanism of a new type is not so much to close the gap between the current prices and the actual production costs, and to eliminate various kinds of imbalances and incongruities, as to set up a fundamentally different pricing mechanism typical of a socialist market. This market has its own distinguishing features, the principal among which should be a systematic government influence on the market's key economic parameters – supply of goods, consumer demand and prices. Thus, reform does not at all mean a shift to an uncontrolled market economy, but requires a certain degree of intervention by the state which, however, will decrease over time.

Under a new economic mechanism, state management bodies should not engage in drawing up detailed plans for production and marketing of goods across the whole range of their desegregated list. Their sole function would be to fulfil state orders for key types of products not on an administrative-command basis, but on the basis of vertical economic relations. State orders derive from key national economic targets and from the guidelines for structural and investment policies laid down in government plans. This fact alone indicates that determining the size and structure of production and supply of basic goods in the transition period of setting up a new economic system should be a responsibility of planning and management bodies of the government.

Now, what goods are to be considered basic, and what requirements should they meet?

I think that the list of basic goods subject to state plan-order should include so-called key or structure-determining goods numbering several hundred items comprising consumer goods as well as means and objects of labor. It is these goods that shape the overall structure of production and consumption in the world economy, and what is essential for the state plan-order is not simply to list categories of desegregated items (for example, shoes, clothes, pipe, metals, coal, ore, etc.), but a list of key goods representing specific items with specific consumer qualities.

The state plan-order placed with individual sectors, associations and enterprises should involve the use of economic regulators with incentives for its fulfilment. Among such regulators, prices play a major role. Therefore the setting of prices for key structure-determining goods is a prerogative of supreme economic management bodies which, when deciding on price levels, take guidance from plan-provided macro-indicators of national economic development, necessary structural shifts, investment policy and the planned situation. At the same time, management bodies examine the impact those prices have on economic performance indices of various sectors, industries, individual associations and enterprises. Prices of key goods included in the state plan-order should serve as additional material incentives for enterprises bidding for that order. A sine qua non is also the joint elaboration of global planning characteristics, target figures and prices of basic goods included in the national economic plan so that a balance is ensured between the total supply and the cost proportions. By and large, this approach can be considered a cross-section of national economy planning where government regulation of the pricing process is the determining factor. At the same time, in the day-to-day economic operation of associations and enterprises that function on the principle of economic independence and maintain horizontal production links between themselves, it is contract relations and, accordingly, contract prices that come to play the decisive part. To the greatest possible extent any uncontrolled developments should be excluded from such ties and relations, not just because enterprises operate under the influence of the state plan and their production programmes are more or less shaped by the need to fulfil state orders, but because horizontal relations themselves take shape only with the use of economic regulators and of the whole code of economic behavior rules established by the state. However, in conditions of the narrow and detailed specialization of many enterprises in the manufacturing sector, monopoly trends, i.e. the diktat of the producer, is very likely to appear inter alia in the setting of contract prices. These processes are rather widespread in the Soviet economy. So, what could be done to combat them? At the initial stage, until such time as economic regulators, in particular a mechanism of progressive taxation of excessive profits, gain foothold, we will be forced to use levers of administrative regulation borrowed from the arsenal of the command economy. For enterprises that continually set inordinately high prices, price ceilings or profitability ceilings in terms of prices should be established, and all profits in excess thereof should be transferred to the state budget. However, at later stages the main role should be played by economic norms, e.g. payments by enterprises for resources they use are to be made at stable rates, valid for 5 years as well as scales of progressive taxes on excessive profits. The latter measure is designed to encourage enterprises to aim at an optimal level of contract prices precluding a desire drastically to raise them since this will have no effect on the economic prosperity of such enterprises anyway.

There is yet another important aspect belonging to a somewhat different dimension – the proportions between the price level and consumer demand. For all practical purposes, the consumer's ability to pay a price depends completely on his finances or, plainly speaking, on whether he has enough money. Today the situation in the production sphere is such that money circulation channels are overflowing mostly with credit money. In bank accounts of many enterprises excess inflation money is accumulating which has no equivalent in a corresponding amount of goods, but at the same time this money enables enterprises to accept the level of contract prices suggested by producers. Therefore a partial reform of credit money circulation would be advisable to take out the excessive money not matched by supply of goods and to improve the financial state of the economy. It may even be sensible to effect through such reform a degree of shortage of consumer demand in the production sphere so as to encourage businesses not only to search actively for customers, but also to reduce prices. This is particularly true of the engineering sector with its peculiar situation where costs are growing rapidly in the process of modernizing the range of goods produced.

On the one hand, scientific and technological progress undoubtedly requires fast structural change in favor of advanced and state-of-the-art equipment and technologies. This is demonstrated by the constantly growing output of numerical control machine tools, machining centers, robots and robotic complexes, microprocessors, flexible production systems, etc. On the other hand, it is in the engineering sector that a situation increasingly emerges where the growth of costs and prices outstrips that of the effect attained through the use of advanced equipment.

In the meantime, the cost-based pricing practice not only fails to expose these negative phenomena, but helps in great measure to disguise them, and in some cases even creates an illusion that instruments and equipment get cheaper. To give an example, let me recall the overall review of wholesale prices in the engineering sector made as of January 1, 1982, as a result of which prices in the renewed lists were somewhat reduced. The reduction applied for the most part to those types of products whose actual manufacturing cost-efficiency by the time of the price review proved above the norm. Yet it was those products that were hastily withdrawn from manufacturing and replaced by new, more costly ones. So there was no real reduction in prices for engineering product customers, rather, the converse was the case and prices continued to grow, as evidenced by the dynamics of average wholesale prices.

As regards the average price dynamics in the engineering sector, it is growing virtually in all industries, although the pace varies. It is fastest in the machine tool, tractor and agricultural machine, car, and electronic industries. What is more, price growth is far from always warranted by a corresponding increment in the practical effect achieved through the use of new technologies. As a result, instead of saving social labor and resources, the national economy suffers direct losses. To give an illustration, let me cite comparative price levels and efficiency rates of agricultural machinery – combine harvesters Niva, Kolos and Don-1500. If we take the price of Niva as equal to 1, the price of Kolos will be 1.36, and that of Don-1500 will be as high as 7.11. At the same time, the ratio of efficiency will be completely different, namely, 1:1.24:2.25.

The practice of setting prices of manufactured goods sold to farmers has long retained in its arsenal the so-called system of two price lists – one of wholesale prices for agriculture and the other of wholesale prices for industry. This implies that many types of manufactured products delivered to farmers had two prices – one, which was, as a rule, higher, was intended for the producer plant, while the other, lower price, was intended for the agricultural consumer. What does this mean?

Above all, this means that such products are inefficient for society because the difference in prices shows precisely the amount of losses related to their production and sale. These losses were covered by subsidies from the budget amounting to several billion roubles a year. Ironically, the economic interests of both producers (industry) and consumers (agriculture) in this case were for the most part met, as producer plants were paid a price for the products that included a normal profit, while consumers (collective farms and state farms) obtained those manufactured products at a lower price that they could turn to advantage. As for losses of the national economy associated with compensation of price differences and totalling several billion roubles, they could be recouped only at the expense of some other, highly efficient industries. So it turned out that preferential prices for industry, which created conditions for its profitable operation, only generated an illusion of efficiency, while in reality the national economy sustained direct losses.

In 1988 the lower price scale of industrial products for agriculture was rescinded and agricultural machinery, equipment, mineral fertilizers, fuel, power, and other things began to be sold to collective and state farms at much higher industry wholesale prices. As a result, agricultural costs have risen dramatically, while purchase prices of agricultural products have not changed so far. In order somehow to offset this rise in costs, purchase prices for produce of low-profit farms and of those operating at a loss were increased correspondingly.

So what was the upshot? In the first place, this was one more triumph for the cost-based methodology of administratively regulated pricing, which aims at setting practically all production costs and is a drain on the national economy.

Secondly, this was a blow at the economically most efficient farms which would get no increase in the purchase price of their produce and, consequently, would not be compensated in any way for accruing costs. Thirdly, this was a graphic demonstration of industry's ability to dictate to agriculture that compounds relations of inequitable exchange and blurs the distinction between good and bad economic performance of individual farms.

The problem, however, is not confined to major deficiencies in the system of price relations between industry and agriculture. The methodology and practice of determining agricultural purchasing prices strictly on the basis of costs are basically flawed, too. Here, more than in any other pricing area, the principle of price individualization through the so-called zonal and intrazonal differentiation is wide-spread. This results in the fact that today in the USSR there exist dozens of prices at various levels for each basic product of animal husbandry and farming, and those prices compensate farmers for actual production costs existing in various zones, regions and even individual farms. Furthermore, in case of low-profit or unprofitable agricultural enterprises, special additions to the farm prices are used in order to improve their financial position.

What problems was this pricing system designed to solve? One problem was how to stimulate farmers' interest in turning out the quantity of products specified in plan directives for farms. And if, for example, farms in the Pskov region were directed by plan to grow wheat, the government purchasing price of wheat was set high enough to compensate them for the costs and to ensure a certain amount of profit. That price, of course, was much higher than the purchase price of wheat set for the Kuban region, Kazakhstan or the Ukraine, yet the question of a socially acceptable maximum wheat price for the national economy as a whole does not even arise here. After all, under such methodology, individual prices are set in accordance with individual costs only because the objective pursued is to enhance agricultural output wherever this may be done.

As a result, prices in general no longer serve as an economic barometer of cost efficiency; they not only fail to be a criterion for assessing what is profitable where, but rather the converse is the case – they distort real assessments of economic efficiency, and the growing of wheat in northern areas may thus prove more profitable than in the south. This is why a fundamental change is so badly needed in methods of fixing purchase prices of agricultural produce.

The point of departure here should be the establishment of a level of purchase prices depending on the consumer prices of the corresponding goods made from raw agricultural products. In this context, the trend of change in purchasing prices would not be determined by the dynamics of agricultural production costs, but by the fluctuation of consumer prices. An attempt to take into account in fixing purchase prices the growing price of industrial products consumed by agriculture, seeking in this way to "trigger" a subsequent change in consumer prices, would be nothing else than a recurrence of a purely cost-orientated approach provoking a wave-like growth of prices and solving no economic problems at all.

In our view it is consumer prices alone that should serve as a point of reference in determining agricultural purchase prices, which means that, a system of consumer (retail) prices should be formed under which a balance would be maintained in the consumer goods market. This may be a step-by-step process, but it should be carried on intensively. In turn, purchasing prices should not only be lower than retail prices, but have to be uniform in basic categories of agricultural produce or, in any case, provide for only a marginal differentiation among larger zones. Only then could purchase prices become a criterion for rational specialization of agricultural production, and comparison of individual costs of farms and the said prices would definitively answer the question about the measure of profitability or non-profitability of individual categories of agricultural products. A substantial differentiation of consumer purchasing prices can only be allowed in regard to products that are sold in provincial, local markets.

So, a pricing system whose underlying principle is individual cost compensation, in our view, is not compatible with the challenges of a radical economic management reform.

Coming back to the illustrative example involving agricultural machinery, only those prices can be considered economically sound that reflect the machinery's actual efficiency and the existing demand for it. Thus they establish the amount of expense socially necessary to produce the machinery. That amount is smaller than physical expense incurred by agricultural machinery plants, which raises the crucial question about the fate of such enterprises.

Until recently, it was simply impossible even to broach this question. After all, Soviet economic theory held that socialism directly introduces social relations based on overall planning under which the economic prosperity of any kind of production is guaranteed by the state. Now the situation has changed. The Law on the State Enterprise (Association) has laid down the principle that the state is not accountable for the liabilities of enterprises, and enterprises are not accountable for the liabilities of the state or other enterprises, organizations or agencies. In these circumstances the question concerning non-efficient industries assumes not just an academic, but strictly practical importance – options range from declaring such enterprises bankrupt and subsequently dismantling and selling their property to developing and implementing a set of measures to rectify their economic situation, for example, by granting long-term credits for technological modernization and reconstruction programmes with a view to an eventual reduction of production costs to the socially necessary level. It is clear, however, that the only thing to be excluded from this range of possible economic solutions is the policy of deliberately camouflaging non-efficient economic performance, whether by means of preferential prices, budget grants or any other economic artifices. Otherwise, if the practice of arranging individual prices for non-efficient industries is maintained, and moreover, if those prices provide for an assured profit, then any talk of switching over to economic methods in management under such circumstances would be a mockery of their very essence and of the entire effort at a radical economic management reform.

Meanwhile, there are reasons to believe that it is this approach that administrative pricing agencies pursue by keeping intact cost-based methods of price setting, instead of doing quite the opposite, as they should, i.e. consistently shaping a system of prices that would promote a balance between demand and supply in the nation's economy; and the question whether all sectors, associations and enterprises will be able to operate on the basis of such prices in conditions of self-financing can be decided only by comparison of production costs with the existing level of balanced prices. This is the gist of a cost-controlling mechanism of pricing.

But what is the practical way of introducing such a mechanism?

As we have pointed out earlier, this could be done, firstly, on the basis of singling out the sphere of centralized pricing of key commodities within the framework of the 5-year economic plan where key macro-indicators are formed and, among other things, material-commodity and cost proportions of the plan are harmonized, and secondly, on the basis of the extensive practice of setting contract prices for all other types of products. And here, the importance of wholesale trade in the materials used for manufacture (means-of-production), which should be the primary function of material and technical supply bodies, can hardly be overestimated.

Responsibilities of a commercial intermediary between the producer and the consumer are not limited to delivering commodities from the former to the latter. Trade in means of production is an indispensable and principal feature of a means-of-production market which is going through the initial stage of development. In the beginning, it is material supply bodies that have to manage the market's key parameters – demand, supply and prices. In contrast to the present system of rationed distribution, they will play a qualitatively different role. A commercial middleman is an effective link between the producer and the consumer: he studies, and to some extent shapes, real consumer demand for means of production, places a corresponding order with the production sector, exerts a direct influence on economic contract prices, and himself operates on a commercial basis and on the principle of self-financing by virtue of commercial profit incorporated in the contract price. Relations of commercial middlemanship in the supplies system are just being established and it is essential here, we think, to play special attention to the sale of means of production that are in short supply. It should be the function of supply agencies in a period of transition to a new economic system to determine effective consumer demand

and supply, promote the establishment of differentiated prices in various spheres of application of hard-to-get products, and thereby shape a rational structure of production and consumption of means of production.

However, while the market of means of production in the Soviet Union is just going through the embryonic stage of development, the same cannot be said about the consumer goods market. To be sure, there too some elements of rationing exist in the sale of a number of foodstuffs and manufactured goods, and yet the main hallmarks of a market can be seen quite clearly so far.

Now, what is the actual situation in the consumer market? Perhaps, its principal distinction today is the lack of balance and a more or less short supply of many goods, and this is a decisive factor determining the current growth of retail prices. The fact that consumer demand exceeds supply bears witness to an inflationary character of the nation's economic development, although with a peculiarity. This peculiarity consists in the fact that prices do not serve as an automatic regulator for matching demand and supply. The practice of administratively regulated pricing makes it possible to exercise rigid control over prices or prime necessities and to maintain their stable level for many years, particularly the prices of traditional commodities with invariable qualities. This concerns, first and foremost, food products whose state retail prices have remained unchanged for decades while the scarcity of these products has persisted and even increased in some cases. Still many economists and sociologists consider this policy correct and regard it as a socialist achievement in our country, ensuring a stable level of consumption of basic foodstuffs and manufactured consumer goods by all sections of the population. One could go along with that provided only there are no shortages in the consumer goods market. However, this is not the case, and therefore the actual structure of consumption in various social groups is substantially deformed. Under these circumstances, the relatively low state retail prices of many food products, notably meat and dairy products, not only fail to fulfil their social function of ensuring equal consumption conditions for people in all income categories, but rather do the opposite thing – in relative terms they worsen the economic situation of low-income groups and improve that of high-income ones.

For example, budget study statistics show that 1 kilogram of meat is 60 percent more expensive for low-income consumers than to people from high-income groups. Why is this so? Largely because high-income groups live for the most part in major industrial, cultural and scientific centers or in areas with hard labor conditions where food products are supplied by the state on preferential terms. In conditions of commodity shortages, i.e. when demand is not met by supply, such a deformed structure of consumption is logical since the sought-after commodity fails into the hands of the consumer who happens to be in the shop when it is put on sale. Other logical consequences of commodity shortages include the declining purchasing power of the rouble, which is demonstrated, inter alia, by the fact that it varies from region to region, as well as speculation, etc.

So, the existing state retail prices of food products which are in chronically short supply no longer solve either economic or social problems – they only aggravate tensions in the consumer goods market, twist the level of real consumption to the benefit of high-income consumer groups and create vast opportunities for the extraction of illegal incomes.

In these circumstances, the proposal of some Soviet economists to postpone the solution of the extremely complicated and painful problem of consumer prices by about three years until the situation in the consumer goods market has been improved is, apparently, motivated solely by social and political considerations. This proposal is evidently predicated on a still widespread illusion that a balanced economy can be built under an administrative-command system of management. However what will most probably happen in the meantime?

Clearly, it would be unrealistic to hope for a solution of the food problem in conditions of growing inflation, both hidden and open, irregularities in money circulation and the falling purchasing power of the rouble. After all, if the administrative-command system of management is not scrapped, the budget and credit growth in the nation's economy will continue, which means that the shortage of goods will persist and, most probably, even get worse with all the ensuing consequences. Inaction in the field of pricing, just like in other areas of economic management, for that matter, could only exacerbate the overall economic situation in the country and stimulate the escalation of crisis.

Theoretically, to achieve a balance between the amount of commodities and the money supply, it would be advisable to pursue a policy aimed, firstly, at a drastic increase in the supply of goods and services, and secondly, at revitalizing money circulation through a set of measures to enhance the rouble's purchasing power and its stability in a fairly full market of goods and services, as well as at bringing prices in line with the actual demand-supply proportions in the USSR economy. Many Soviet economists pin their hopes primarily on the possibility of resolving the crisis through a rapid build-up of consumer goods supply, assuming that this could be done by intensively restructuring the economy, converting military production, scaling down centralized capital investments in the production sphere and reducing the number of incompletely constructed facilities and of uninstalled equipment, and expanding the volume of consumer goods imported.

Such measures may attain some effect, but only a marginal one and in a rather distant future, except for imported consumer goods, because the intention is to enforce them in the framework of the same old cost-oriented and ruinously inefficient administrative-command system. A quaint logic is at play here – first the market should be brought into balance, and only then should reforms be launched. But if the former task could be accomplished under the present economic system, why initiate reforms in the first place? However, if commodity shortages and inflation within the economic system currently existing in the USSR cannot be reversed in principle, then any delay of a radical economic reform is a road to a total collapse of the economy.

We think that shortages and inflation are reproduced objectively and constantly by the administrative-command system itself, they are part and parcel of it, and therefore can be eliminated only if the entire system is dismantled. However, a change-over to planned market relations, being a very difficult thing to do, presupposes the ability to manage the economy using such tools as money and prices. One is bound to note in this connection that the declared tasks of radical economic reform include the need not only to change fundamentally the system of prices and pricing, but also to make a cardinal transformation of the financial and crediting mechanism. This is yet another proof that all instruments of market regulation can only operate effectively when used in close coordination.

While price reform guidelines have been recently the subject of a broad, extensive and even detailed discussion in the USSR, measures to improve the financial and credit situation of the national economy have been defined in terms of traditional approaches, in a very generalized form, with no clear schedule, stages or end results. In the meantime a number of Soviet economists believe that without a monetary reform it is impossible to normalize money circulation in the country and to make a decisive breakthrough in revitalizing the financial system and enhancing the stability and value of Soviet currency. And in the final analysis, a price reform itself can be effective if coupled with monetary reform.

It should be said that all monetary reforms have some common features and some individual distinctions, pursue strictly defined objectives and are carried out in a historically specific situation. Soviet history has seen two major monetary reforms (in 1922–1924, and 1947) which had an important feature in common – both were designed to help to normalize money circulation utterly disorganized by wars. There also was a major difference: the monetary reform of 1922–1924 was intended not just to normalize money circulation in the country by introducing a new currency unit (the gold chervonets) and gradually removing from the circulation the old money (called "sovznaks"), but also to give the new currency a gold value and make it convertible. It would be worthwhile to recall that the policy of the time aimed at integrating the USSR as much as possible into the international division of labor and at the development of cooperation of all kinds with foreign countries. To do that without a stable and convertible currency would be very difficult.

The monetary reform of 1947 had, as a matter of fact, no such objective because the rouble's parity with the US dollar was quite intentionally established at a level 3 to 4 times higher than the real purchasing power of the Soviet currency at that time. The declared gold content of the rouble (1 rouble = 0.22 grams of gold) was purely nominal as there was no actual exchange of roubles for gold. During that time, or, to be more specific, in the early 1930s, the rouble lost its link to gold and turned into a regular form of credit money serving the purpose of payment settlement. Its nominal value could only be supported by a possibility to convert it into goods and services in the process of money circulation. However, the rouble's exchange rate in relation to other currencies, as calculated by the Gosbank of the USSR, continues to be a far cry from its real purchasing power, and due to this the possibility of making it convertible is linked to the need for a devaluation, and a devaluation not by tens but by hundreds of percentage points.

Some steps to that end have already been taken. The Decree adopted by the USSR Council of Ministers at the end of 1988 "On the Further Development of Foreign Economic Activities of State-Owned, Cooperative and Other Public Enterprises, Associations and Organizations" stipulated that as of January 1, 1990, a 100-percent increase be made in hard currency exchange rates with respect to the rouble, i.e. the rouble was actually devalued and its exchange rate was cut by half. Besides, as of November 1, 1989, the rouble's exchange rate with respect to convertible currencies was slashed by a factor of 10 for non-commercial payments, which should result in a dramatic decrease of currency speculation inside the USSR and stimulate the flow of foreign tourists to our country. At the same time, presumably, there will be an increased strain on the domestic market of foodstuffs, services and some industrial products leading to additional tensions in the market, tight as it is.

However this is only one side to the problem. Another one, which is no less important, concerns the questions of domestic money circulation proper, analysis of the negative trends that already exist there and the search for ways to reverse them. Disorder in money circulation today is less severe than right after the war, but still it is considerable and continues to grow. The enormous state budget deficit requires substantial issues of both cash money and credits, and incidentally, the latter amounted until recently to tens of billions of roubles a year. As a result of the influx of excess money in to circulation, the purchasing power of the rouble has fallen noticeably, money deposits in people's bank accounts have increased dramatically, and at the same time consumer goods in the market have become more scarce.

Moreover, people have tens of billions of "hot" cash money in their hands which can be immediately used to buy the sought-after goods as soon as they appear. This stimulates a situation of feverish demand and creates a generally explosive situation in the consumer goods market. Retail prices have been growing at the annual rate of 5 to 6 percent which, combined with inflation of money, provides for the annual inflation rate of 10 to 11 percent. These crisis phenomena in the Soviet economy did not appear overnight. Way back, in the wake of the 1947 reform the volume of cash and credit money turnover gradually started to increase, and in parallel a commodity-price imbalance began to develop in the USSR economy. A number of factors were at play here.

Firstly, the amount of produced goods and services kept increasing, which called for a corresponding growth in the amount of money in circulation.

Secondly, reviews of wholesale prices of industrial products in 1949 and 1955 and the intensive growth of purchasing prices of agricultural products, beginning in 1953, raised the estimated value of gross national product faster than the pace at which its useful effect increased in the consumption sphere.

Thirdly, a relatively high growth rate of wages and labor remuneration in all spheres of material production, and above all in agriculture, was not accompanied by a corresponding increase in the production of consumer goods and services, which resulted in the declining value of money earnings and a decreased purchasing power of the rouble. In other words, production was, and continues to be, insufficiently responsive to rising prices.

Fourthly, an exceedingly high growth rate of industries in Group 1 of the national economy and the ever increasing involvement of production and labor resources therein were accompanied by a relatively slow development of Group 2 industries with growing disproportions and an ever greater tilt of the economy towards production for production's sake.

The result was that by the beginning of the 1960s the money circulation system in the country was functioning with great difficulty, and hidden inflation ever more frequently manifested itself in a fairly open form of direct increases in wholesale and retail prices of individual goods and in the growing scarcity of production assets and consumer goods and services.

In 1961, with a view to improve the financial situation at least partly, a limited money reform was undertaken involving the denomination of the rouble in proportion of 10:1 and the introduction of new paper money to replace the old rouble bills. Because the same ratio was used to change the price scale with a certain approximation of prices of individual goods and services, not divisible by 10, to the benefit of the state, it can be assumed that Gosbank of the USSR did derive a certain financial gain out of the whole business. But on the whole, that measure fell short of normalizing money circulation in the country because the causes of the unjustifiable "pumping" of money into circulation were not addressed. An attempt in 1965 to enact an economic reform intended to dismantle the administrative-command system of management was a failure. And since the command system of management survived, all budget and credit channels for putting inflation money in circulation remained intact as well.

What is more, this process intensified in the following years and at some stage even became hard to control. The growing government spending, particularly defence outlays, could not be fully provided for by proceeds from public production. Therefore the government was increasingly forced to appropriate for them using Gosbank credits, which in the final analysis turned out to be subsidy grants, and the outward appearance of a deficit-free budget increasingly came into conflict with its actual status. Under the 1989 plan, the direct state budget deficit amounted to 36.4 billion roubles plus 63.4 billion roubles borrowed from Gosbank as money of the so-called Nationwide Credit Fund. Thus, the sum total was about 100 billion roubles. In actual fact, however, the government announced as early as mid-1989 that the real budget deficit would reach 120 billion roubles, or 14 to 15 percent of the value of GNP. This brought the country to the brink of economic catastrophe.

The huge budget deficit predetermines an intensive pumping of inflation money into circulation, which, no doubt, will affect the dynamics of prices in the next few years. Even if tough administrative measures are adopted to check the price growth – and this is exactly what the government seeks to do now – it is no longer possible to stop it. It is most likely that the hidden price growth will accelerate in the next few years, which, of course, will help to some extent to "tie up" surplus money through the sale of goods and even to expand budget incomes, but those will be fictitious incomes with no real consumer goods or services behind them.

To be sure, there has existed, and in some measure continues to exist, a major source for supplementing the Soviet budget incomes with real money, as proceeds from foreign trade, primarily from the sale of oil, natural gas, and timber at fairly high world prices, particularly in the early 1970s. However, this factor only delayed the financial and money circulation crisis.

What kind of proposals are being made now to achieve a radical improvement in the financial sphere of the USSR national economy taking into account its actual status and conditions of economic reform?

There are two alternative approaches that enjoy widest support. The first one provides for an overall monetary reform with either immediate or gradual removal of surplus inflation money from circulation. Under the second approach, a monetary reform is not advisable, but a tough programme of transformations in the financial and credit sphere should be developed and implemented which, in the long run, would contribute to the achievement of the same result – a balance between solvent social demand and supply of goods and services in the national economy.

If the first approach is to be tried as the most radical means of financial recovery, the objectives and possible outcome of a monetary reform need to be carefully juxtaposed. The objective is to normalize money circulation by removing excess means of payment not matched by a corresponding amount of available goods from the turnover of cash and credit money. As regards enterprises and associations, a portion of their funds can be impounded upon recommendation of competent commissions comprising representatives of production units and the financial and crediting system. Yet it should be borne in mind that the state's confiscating or simply writing-off even a portion of the 250 billion roubles amassed by mid-1989 in bank accounts of enterprises and associations would be tantamount to a complete break with the declared principle of economic reform and a return to the rigidly centralized administrative-command model of economy. Still some way has to be found to counterbalance the budget deficit of 60 billion roubles planned for 1990. In this situation the USSR Ministry of Finance, in proclaiming a new approach to financial planning, claims that it will no longer resort to direct and non-repayable Gosbank loans from a special credit fund to redress the deficit. Another solution has been found under which the surplus money is to be taken from enterprises, though not by forcing them to transfer it directly to the budget, but in the "modern" from of mandatory purchase of state-loan bonds to the total amount of the budget deficit. The result would be that a substantial share of the funds of enterprises would be transferred to the budget, and yet this will not by any means solve the deficit problem, because the funds thus expropriated would, in fact, be "empty" as not matched by a corresponding amount of available goods, and for this reason they should simply be excluded from circulation.

In the case of the population, the task of raising the purchasing power of the rouble as a result of money reform can be accomplished through depreciation of money loaned to the government or stashed away in "cookie jars". In practical terms this means that subject to partial depreciation would be people's money in excess of a specified minimum and kept in a saving bank or at home.

Advocates of such reform believe that in this way it will be possible at one fell swoop to do away with underground millionaires and depreciate tens of billions of illegal roubles. Yet it is perfectly clear that there can be no scientific rationale for determining a ceiling for savings to be exchanged under the reform at the rate of 1:1, with what is in excess thereof to be exchanged at the rate of, say, 1:10 or 1:100. Whether the ceiling is set at 5.000 to 10.000 roubles, its voluntaristic nature is obvious and therefore such a "surgical operation" will inevitably affect the honestly earned and accumulated money of working people in many social categories. Tensions run high in the Soviet society as it is, and they need not be raised any further.

But perhaps the monetary reform could succeed in turning underground millionaires into beggars? Hardly so, because the bulk of their money has been either converted into jewelry, immovable property and other reliable means of realizing illegal incomes, or "laundered" through cooperatives and the other types of business ventures now permitted.

This leads us to the conclusion that a one-step confiscatory money reform would fail to bring positive results, and therefore it is the second alternative approach that should be implemented – a consistent series of measures to revitalize the entire financial and crediting system in the USSR encompassing monetary relations in both production and consumption spheres.

It is common knowledge that the gap between solvent demand and supply has been widening in recent time not only in the case of consumer goods and services, but also in the case of means of production. Disproportion there, just like in the consumer goods market, is both general and structural in character, however its scope is hard to ascertain since there are virtually no open statistics available. Because there is a shortage of goods, immediate switch-over to a free trade in means of production would mean a situation where everything or almost everything would be sold out in a jiffy, which would paralyze the economy and necessitate a quick return to the system of rationed distribution of means of production. No doubt, this is a serious argument that cannot be disregarded. But how could market shortages be eliminated? Probably, only by exerting a purposeful influence on solvent demand and supply, and a combined influence at that, because there still is a wide-spread illusion that it is somehow possible to freeze the enterprises' solvent demand for means of production and at the same time keep boosting their production and supply, thus alleviating the shortages. This cannot be done, among other reasons, because every enterprise is both a producer and consumer at the same time, and to produce something it needs resources, so additional supply necessarily gives rise to extra demand.

It is inflation money kept in the bank accounts of many enterprises and organizations which has no equivalent in available goods now nor at any time in the near future, and which, presumably, could be most easily written off or devalued step by step. However, the situation is unique in that money received by enterprises in exchange for their products or services in a real transaction is not inflation money; it becomes that in conditions of a general imbalance in the national economy and as a result of Gosbank's money supply and credit policies, which, of course, cannot be influenced by any individual enterprise. To be sure, it may be relatively easy to pinpoint the inflation-related character of some credits, but the arguments adduced to justify the very need for them are such that being in the shoes of enterprises or industries it is sometimes very hard to question them.

The upshot is the continuously reproduced imbalance in the national economy between the amount of commodities and prices, as well as chronic shortages that tend to get worse. In view of many Soviet economists, the problem could be solved, on the one hand, by expanding production and supply of goods and services, particularly those which are in very short supply, and on the other hand by limiting solvent demand for production resources, as well as for products and services consumed by people in everyday life.

Naturally, the question suggests itself: how could this be done? After all, the problem of a growing imbalance between demand and supply in the Soviet economy has been no secret at all for many years now. Declarations have constantly been made that it needs to be resolved without delay through increased output practically in all sectors of production. And the output was, indeed, on the rise, although its pace was falling with every year. Even today many economists still believe that if the pace were to be accelerated, 2 to 3 five-year periods would suffice to bring the national economy back into balance at last. I think this is a profound delusion.

The economic programme for 1986–1990 contained two contradictory and, in effect, incompatible requirements – a stepped-up socio-economic growth and a radical economic reform. After all, if a radical economic reform is on the agenda, this means the dismantling of the old economic system and the creation of a fundamentally new one. In turn, the new system is supposed to break decades-long economic relations that constitute the basis of society. And such a breakup cannot be accompanied by accelerated social and economic growth, even less so because it also requires a dramatic restructuring of the economy to the benefit of industries which directly specialize in meeting people's needs.

The problem, however, is not that it has proved impossible to achieve the planned increase in the economic growth rate, but that the economic reform from the very outset has been conducted in an irresolute and inconsistent manner, as a result of which the old economic mechanism of the command economy continues to function, although with occasional glitches. And an economy built on the supremacy of barter-type and proportions and managed by administrative-command methods is by its very nature shortage-prone.

The relative surfeit of money in cash and credit turnover is caused, above all, by the fact that funds are appropriated according to what is planned in physical terms rather than according to actual results; there will always be surplus money because even in the process of fulfilling an ideal plan upsets occur, such as product losses and thefts, withdrawal of money from circulation, manufacture of useless or outright defective articles; and as regards manufactured consumer goods, many of them are in no demand whatsoever and accumulate in non-liquid stockpiles. At the same time, supply of money for the public production is guaranteed, so there always is some amount of excess funds which finds no equivalent in commodities. As the economy grows in scope, that amount automatically increases, thus becomes greater with every passing year, which is attested to, among other thing, by the accruing state budget deficit.

The situation is constantly getting worse owing to the fact that a huge share of resources is deflected to the construction complex, where the pay off has a long time lag, to the defence industry, as well as to the support of inefficient enterprises. Excessive spending in those spheres contributes to the exacerbation of the overall imbalance in the national economy.

So, the attempt to materialize the concept of accelerated development through a build-up of production and supply of goods and services in the framework of the present economic system has not attained, nor can it attain, the sought-after goal. This concept derives from production, and presents, in essence, a theory of extensive and cost-oriented growth, and does not include such a major parameter of economic development, such as real solvent demand on earned rather than artificially calculated incomes. It proceeds from the premise that all that has been produced will be sold, and consequently, what is produced should be consumed. There are virtually no limits for this sort of consumption, since the solvency of any consumer is assured even before the product in question is manufactured. Actually, this is the only way that an economy based on the barter-type exchanges of material goods can operate, for money here follows the commodity and performs, as a matter of fact, only one function – that of continuously serving the production process, i.e. manufacturing and distributing "material goods". This is why there is always enough money in the national economy, and what is more, it turns out that the amount of available money, as we have noted earlier, considerably exceeds what is necessary to ensure an effective reproduction process. Hence the conclusion that the scarcity of products is an inevitable and incorrigible state of an administrative-command economy.

The government borrows money to eliminate, at least partly, shortages of goods, but borrows from its own state bank, so there is no need ever to return anything. If money is issued expressly to offset the difference between budget expenditures and incomes, that is patently inflation money for it has no equivalent in commodities. For example, if, after the adoption of a 5-year plan where everything should be balanced, at least on paper, an additional sum of several billion roubles is allocated for social development needs, there will be no additional material resources available to be bought with that money, so in order to make use of the additional funds, resources would have to be taken from other sectors. But that would provoke commodity shortages in those sectors.

Thus the elementary truth is borne out that the pumping of excess money into circulation can only aggravate the economic problems of a transition period, including, of course, the problem of shortages. But are there any ways to solve these problems, and is it possible quickly to achieve an economic balance through some kind of drastic measure?

In principle, such ways do exist, and the central among them is a radical restructuring of the USSR national economy. What does this mean? First and foremost – a basic shift from industries in the fuel, power and raw material sector to manufacturing industries whose end products include consumer goods and services.

It would be an illusion to hope for an early accomplishment of this task, and there are more reasons for that than just the inertia of the economy and the bureaucratic clout of raw material ministries. The main driving force of the present structure of management is the cost-oriented economic mechanism which has no effective regulator to constrain the ever-growing demand for material resources on the part of industries, associations and enterprise. As we have noted earlier, it is only the money actually earned that can serve as such a regulator, and any loss-reporting sector, association or enterprise not only should be unable to expand production, but rather should curtail it. Such is the law of the market. However this is not the case in a command economy, due to the failure-proof operation of the mechanism of budget subsidies and credits which eventually prove unrepayable, i.e. turn out to be the same kind of subsidies.

Some believe that prices are to blame for all this. Can it really be so, they ask, that coal and timber industries operate at a loss, and the electric power sector is several times less cost-efficient than electronic engineering or the shoe industry? Does work there require less skill and knowledge, and is it all that inefficient or inadequately efficient?

Hence the conclusion: prices are bad and need to be adjusted. And since the criterion for determining the correctness of prices is their conformity to actual production costs, a theoretical justification of administratively regulated pricing becomes fully logical. Because individual sectors, associations and enterprises have found their place in the existing structure of production and the plans that are drawn up for them are predicated on the achieved level of production of goods in question, it would only be logical to demand that prices be set under which production of any kind would be more or less equally profitable. It is no accident that in all cost-oriented pricing models the central and most important element is the calculation of a common normative profitability to be included in prices.

In the 1930–1950s, wholesale prices in industry took into account the so-called minimal normative ratio of profitability to prime cost which amounted to 3 to 5 percent. In the 1960–1980s, in connection with the attempted economic reforms this norm was set, as a rule, at the level of 15 percent in relation to the cost of fixed and circulating production assets. And on the occasion of each single price review the profitability of individual sectors and of a majority of enterprises reaches to a significant extent a common level. However, later on happens what should have happened – production conditions, including the cost of production, change differently in various sectors; where the cost is rising the profitability is decreasing and at some point can fall below zero, while in sectors where the cost drops off, the profitability increases. When this differentiation of profitability, in the view of the planning authorities, transcends the limits of the allowable, a new price review is made and everything starts all over again.

It is indicative that the declared objectives of all massive single price reviews in 1949, 1955, 1967, 1982–1984, and 1988–1989 were: first, to bring prices as close as possible to the socially necessary labor input, second, to strengthen the stimulating role of prices in resource saving and in the frugal use of raw materials, fuel and energy; third, to encourage through prices scientific and technological progress, raise the quality of products, promote progressive changes in the structure of production and consumption and so on. However, none of the said objectives has been achieved over the last 40 years. What is more, negative trends have been growing in the economy which have resulted in the diminishing effectiveness of production, a drop in many key social and economic parameters of development, and the state of stagnation and crisis.

It is clear that attempts to preserve the mechanism of administratively regulated pricing and the entire system of economic management can only add to the crisis. In the meantime, the price reform that has been prepared by the USSR State Committee for Prices and which is pressed for by the Ministry of Finance and Gosplan (State Planning Committee) of the USSR, in principle, repeats the massive price reviews of the past decades and does not by any means correspond to the basic ideas of a radical economic reform. Even if it proves possible to redress through such reform the so-called discrepancies in price levels and disproportions between individual sectors of the economy and product categories and types, this will not last long and everything will return onto its circuits after some time, and not merely return but give rise to new and more complicated problems. After all, a price review now would affect not only prices and tariffs in industry, agriculture, construction sector and transport, but retail prices as well. Since the whole spiral starts in full and raw material industries where prices are planned to be raised by 50 to 100 percent, there will be cost rise in all sectors totalling about 200–250 billion roubles. Calculated in relation to the annual volume of GNP it will amount to 13 to 16 percent.

And what does an increase in the estimated value of gross national product by tens of billions of roubles mean if it is known in advance that this will not be accompanied by an adequate increment in the material-commodity content of GNP?

Firstly, this means a sharp rise in the amount of cash and credit money in circulation, and a further decline in the purchasing power of the rouble.

Secondly, a greater imbalance between the amount of available material commodities and money in circulation, since an administrative raising of prices without a concurrent alteration of pricing mechanism itself and the entire system of management of the basis of economic methods cannot solve any of the outstanding and very complicated problems of the national economy.

It should be said that the cost-oriented fly-wheel of administrative pricing is bound to comply with certain objective conditions which appear at a time when the size of subsidies to support prices of a portion of end produce nears the size of total profits derived from the rest of the produce. In other words, if the balance between incomes and losses associated with the end product reaches zero, the economy as a whole "gets stuck", it loses the financial sources needed for expanded reproduction, and the whole process of production and distribution of consumer goods and services lapses into a state of permanent stagnation.

At present, the proportion between the size of price subsidies and the turnover tax incorporated in retail prices has come close to 100 percent and still continues to grow. This process will be spurred by the scheduled reform of wholesale and retail prices whose planned increase will lead, on the one hand, to substantial reduction of profits and turnover tax in the current prices of manufactured goods, and on the other hand, to a further expansion of subsidies to support food prices. As a result, the sum total of subsidies would exceed by tens of billions of roubles the size of the turnover tax, something that predetermines the need for a major one-step increase in retail prices.

What will the consequences of such a price reform be?

At first glance, financial implications of the reform during the initial stage of the existence of new prices would be quite favorable, since there would be a rise in the estimated value of the volume of social production, the profitability of some sectors and industries would grow, the unprofitability of many types of products would be redressed, and the conditions needed to introduce self-financing would be created. However, the economic process is not something static. An overall rise in prices accompanied by a simultaneous overall increase in people's incomes (without which there can be no rising consumer prices due to socio-political reasons!), would as early as the next reproduction cycle inevitably lead to a cost growth in all spheres of material production, and all the current problems of the Soviet economy would simply be elevated to a higher price level. After all, this has been the outcome of the cost-oriented administrative price reforms of the last four decades. But this time the results may come considerably sooner and in a more destabilizing way.

So the review of all prices and tariffs prepared by the administrative-command pricing bodies cannot yield any positive results and, in our view, it must be stopped. But what then is the proper way of conducting a general reform of prices and pricing, what are the goals to be pursued, and what should be done to achieve them?

I think that, as a matter of priority, a new pricing mechanism should be created which would be part and parcel of a socialist market subject to systematic influence and regulation by the state. Therefore, the strategic objective of the economic development should be the establishment of full-blown market relations in the Soviet economy, i.e. markets of goods, labor, money, bonds and shares. These relations are in an incipient phase now. In fact, it can be said that only a consumer goods market, incorporating a considerable element of rationing system, and an even less developed market of labor resources and means of production exist today. However since the logic of market relations is now most felt in the consumer goods and services sector, it is above all there that a set of measures should be implemented to revitalize the economy, as envisaged by the economic reform.

These measures should be carried out in relation to all elements of the market – supply, demand and prices. Here, illusions should be discarded that a balance in the market could be achieved by freezing prices and only building up supply or by restricting in some way consumer demand. A market can only be called a market when all its elements interact dynamically and are in constant motion. In addition, and we have pointed it out before, all our previous attempts to fill the market with goods in conditions of rigid administrative regulation of prices only lead to further inflation and, in the final analysis, to a greater imbalance between goods and money. If we are to develop an anti-inflation programme now, we obviously should not predicate it on a tough administrative regulation of prices of millions of items. At the same time, the condition of the consumer market is such that a further price growth – either open or hidden – is inevitable. Which option should be preferred and could this process be controlled?

In our view, centralized management of overall demand, supply and prices in a socialist economy is not only possible, but indispensable. However, it should be totally different from what we have now. Because state planning and pricing can only encompass a few hundred basic goods and services, it is only to these product categories that the state order should apply, and prices and other economic standards would be established which would stimulate enterprises to fulfil the order. As for millions of all other commodity items, they should be made subject to contract pricing between producers and consumers.

The first, and perhaps the main, problem constantly confronting the national economy is the obvious resistance of ministries through the instigation of price growth. All attempts to stop this process by administrative methods do not bring any practical results, since in a situation of global shortages the producer "goes for the jugular" of the consumer trying to extract from him consent to the most inflated prices as compared with production costs by threatening not to conclude or to terminate the contract. Could economic methods be employed in this situation? I think, they could, and the decisive role here is to be played by the tax system combined with subsidies to consumers who are being robbed by producers through prices.

Since the state establishes economic norms in a centralized fashion, they should include progressive tax scales, which at the initial stage should be differentiated for various sectors and, perhaps, for some categories of products. Later on, as the financial situation of the country improves the scope of the differentiation could be drastically narrowed.

Financial and credit policies in conditions of an unbalanced economy need to be supplemented with a corresponding incomes policy. Naturally, when prices are raised, incomes grow too, but it is one thing when this involves the production sphere and people employed there, and it a completely different matter when these processes affect non-production spheres and people with fixed incomes. A market pricing mechanism can only be used when there is flexible regulation of people's incomes.

So, a reform of the price system and of the pricing mechanism, in principle, should be started with the development and nationwide discussion of various options for an initial change of prices of basic goods and services in the consumer basket. In parallel, the problem of the necessary reform of the entire system of people's money incomes would be tackled, or as a palliative, the introduction of compensatory supplements to the existing incomes. A list would be drawn up of key goods and services whose prices would stay unchanged during the 5-year period and would be directly controlled by the government. Prices of all other goods and services may fluctuate depending on the interplay between demand and supply.

As for a reform of wholesale prices of industrial goods and purchasing prices of agricultural products, the same mechanism of balanced pricing should be applied there too, but the starting impulse could be provided by simultaneous review of key product prices seeking to achieve maximum harmony between the material content and price structure of the gross social product.

This approach to the reform of the pricing system could be have been realized as early as 1987, which we actually proposed. However time has been lost. The economic situation in the USSR is such that the consumer market is almost in a shambles, and the negative trends continue to worsen. The expanding scope of shortages and inflation push the nation to the brink of a financial collapse, which could make it necessary to introduce an overall rationing of basic foodstuffs and manufactured consumer goods. The possibility of continuing the economic reform could survive only if, in addition to the rationing system, a broad network of shops is set up where prices would determined by a balance between demand and supply. In this case, chances would still remain for shifting over time to market pricing in all spheres and phasing out the rationing system.

It becomes increasingly clear now that the current economic reform has so far failed to solve the central issue of giving real economic independence to basic units of production and has come up against a very serious obstacle – the impossibility of shifting overnight to market relations and real prices associated with them. However even extraordinary measures in the economic field can be implemented from different "ideological" positions advocating either a return to the old ways or the quest for avenues of transition to a new system. I dare say that only the creation of conditions for an effectively functioning market can in due course extricate the economy from the crises into which it has been led by the administrative-command system of management. The road to that goal will be long and difficult. Yet there is no alternative and we must follow this road consistently and unswervingly.

Chapter X

Attempts to adapt the managerial bureaucracy of the USSR to a market economy

Igor S. Oleinik

PhD Economics, Senior Researcher, Institute of Economics of the Academy of Sciences of the USSR

Over the last few years the Soviet leadership has taken the first steps away from rigid ideological dogmatism towards pragmatism. Even to the Communist Party's elite, it had become completely clear by the middle 1980s that the traditional economic system based on "administrative–bureaucratic socialism" was doomed, even if somehow it were possible to revert to the system which operated in Stalin's time. Therefore, the Soviet ruling elite had no real choice other than to initiate reforms, particularly in economics, simply in order to survive. At the start of the Perestroika period a few of the senior CPSU leaders including M. Gorbachev thought that a simple, and limited, renovation of the archaic and practically defunct "ship of state" would be possible, and that the rate of industrial development could be increased without radical changes in the political system, in the prevailing ideology, or in administration and management. However, real life today is nothing like that which the social system created in the USSR 70 years ago was designed to foster; indeed, that system could itself only be kept in being through the constant, and sometimes very cruel, suppression of economic and political freedom.

After three years of Perestroika, it has become clear to most people in the Soviet Union that the economy cannot be modernized and made efficient without radical reform of the political and economic systems. By the end of the 1980s the new concept of "radical economic reform" had been gradually transformed into a new concept, namely, the creation of the "planned-market economy". Consequently, after about five years of futile attempts to organize another "great leap forwards" in Soviet society, one of the main dogmas, the concept of centralized planning within a one-party state had been, for all practical purposes, abandoned.

Already during the first stages of "radical economic reform", 1985–1989, it had become obvious that fundamental changes in Soviet society were being obstructed by many difficulties. First and foremost should be mentioned the widening gap between the reforming faction at the head of the CPSU, and the politicians forming the middle of the party hierarchy who, together with the senior industrial managers saw these reforms as a direct and personal threat to their power and prosperity.

The basic contradiction in Soviet economic management today is therefore that the state and industrial bureaucrats do not really want to put into effect the "planned-market economy" that senior politicians want, but neither do the latter really feel ready to reconstruct the bureaucracy and set free the army of potential entrepreneurs awaiting their change to work under real market conditions. Consequently what had been a clear and well defined strategy was put into effect in a half-hearted, contradictory and inefficient way. The new laws which were to form the legal basis of economic reform were, it is now clear, badly drafted with many poorly defined statutes which have delivered into the hands of the bureaucrats the means to thwart all really radical changes in economic practice.

The experience of the last few years has made it abundantly clear that all attempts to take steps to create a market economy in the USSR have met enormous resistance at every stage from every level of the bureaucratic hierarchy. This bureaucracy has become far too big and powerful and the experience of Perestroika has clearly identified it as one of the biggest barriers to the introduction of a new economic system.

How has this come about? Why do upper and middle state and enterprise managements put up such resistance to reform? Can the major opponents of reform, in the State, Party and industrial hierarchies, be identified? Can the future course of events be predicted? How can this bureaucracy be neutralized? To visualize the general situation today and to seek answers to these and many other questions, it is necessary first to describe the ways in which this bureaucracy has emerged in the USSR over the last few decades.

1. The Nature of the USSR Bureaucracy and the Processes of its Development

Very many times in the history of the Soviet state, the governmental and managerial machinery which controls the clumsy and rusty socialist economy has been modified. Each modification was introduced so that the machinery would be able to cope with a new set of tasks imposed on it by communist politicians in each successive stage of the historical development of the first socialist state. Though the stimuli prompting these modifications were different, there was no change in how personnel were selected and allocated to different jobs. The aim was always to strengthen the bureaucracy so it could exercise centralized control, leadership and management of all processes in the economy.

The proliferation of this bureaucracy, and its penetration into all stages of the management process in all sectors of the economy and the state, was made possible by the fact of complete state ownership of land and of the means of production, and because state control of material and even of labor resources was nearly absolute. For many decades the main function of the socialist bureaucracy was the control, through the directive planning system, of the programme of socialist industrialization initiated under Stalin. Industrialization caused the number of economic units to increase dramatically year by year. These units themselves grew bigger and the technological and economic connections among them became more and more complicated. Centralized control of all these units necessitated concomitant growth in the number of bureaucrats throughout the hierarchy. About every ten to fifteen years the government tried to prune this bureaucracy but these efforts always failed and the bureaucrats continued their growth in numbers and in power.

Even during the recent period of "stagnation" the bureaucracy of the Soviet Union continued its remorseless growth. In 1970 there were 13.6 million people involved directly or indirectly in the management of the economy – in 1975, 15 million; in 1980, 17.2 million; and in 1987, almost 18 million. This last figure represents about 15% of the labor force of the USSR.

Over the last three years yet another campaign to curb the bureaucracy has been running, launched as usual from the very top of the government and from the very heart of the Party machine, and indeed the total number of bureaucrats has been slightly reduced. But there has been no real change in the general structure.

The position at the end of the 1980s was therefore that about 15 million people were employed in management and administration of the production and distribution sectors. Some 2 million people were employed directly by the State in ministries and suchlike. A further 350–400,000 worked in the administration of the Party, trade unions, cooperatives and other governmental and quasi-governmental organizations. Finally, 4 million people provided services for the higher echelons of the bureaucracy, as drivers, guards, couriers, secretaries and so forth. [1]

At this time, the USSR had about 100 All-Union ministries and state committees of which more than half were involved in direct leadership of economic and financial activity, employing more than 100,000 officials.

For several decades new ministries have been created simply as a result of top-level administrative decisions. Many of them are now very powerful monopolistic organizations founded on principles of total autocracy. There seems to be no standardization of function among ministries of the republics and the All-Union ministries even though they may bear the same name, and many of

---

[1] Pravda, 21 January 1988.

these ministries simply duplicate each other. There are great disparities, too, in the sizes of the economic sectors controlled by different ministries. Some such sectors may have a volume of production up to 52 times greater than others, may employ up to 55 times as many staff, and may involve 74 times as many subordinate enterprises, and so on.

Considering the Union Republics, in 1989 there were more than 800 Union-republic and republic ministries and committees staffed by more than 135,000 people. Generally, there was no relationship between the number of ministries and committees and the sizes of the industries involved. For example in very recent times the number of ministries and state committees in such republics as Kazakhstan, Armenia and Azerbaijan was greater than in the Russian Federation though the industrial production of the latter was hundreds of times bigger and its land area much larger. In the Moldavian Republic 53 ministries supervise 78,000 independent businesses. In the Russian Federation almost the same number of ministries controls 273,000 businesses – 35 times as many. There are many other examples. Indeed, bureaucrats are, numerically, particularly prominent in the Turkmenian Republic, Kirghizia, Latvia and some other republics. Throughout, there are to be found the same ministries at republic level as there are at All-Union level. Further, in several republics, subdivision of administrative areas into unjustifiably small units has taken place.

Before the Revolution, Russia, with its 70 provinces, was considered excessively subdivided. After 1918 these were merged into 28 provinces. However, the process was soon reversed; during the operation of the command-economy system in the USSR there were created 15 republics, 129 territories and 20 Autonomous National Provinces each with a typical bureaucratic government. These bureaucracies alone employed over a million people. [2]

It has been estimated that about 14% of all the national income of the USSR is spent on its bureaucracy. In developed countries the figure is between 7 and 10%.

Every five or ten years, the top levels of the hierarchy attempt to prune the lower echelons by a small percentage but the effects of these attempts are never long lasting.

Many studies of the ways power is held and exercised in the state have found that the extent of bureaucratization of the state machine depends directly on how authoritarian is the political regime. As a consequence of this dependence, bureaucracy results in "excessive separation and alienation of the state machine from society, growth of egocentricity of officials, use of the power vested in them by the state to further the interests of the bureaucrats as a group rather than to advance the interests of the whole of society". [3]

---

[2] The USSR has a far higher proportion of bureaucrats than other countries – 10% of the labor force. Compare the People's Republic of China, for instance – 5%.

[3] B.P. Kurashwily, Struggle against bureaucratism, Moscow, 1988, pp. 8–9 (in Russian).

In order better to understand the development of the state bureaucratic machine which still functions in the USSR, it is necessary to consider its history.

The basic components of this system are the industrial ministries. These were created at the beginning of the 1930s when the concept of the "industrial leap" was adopted, that is, the strategy of accelerated industrialization. But it is important to stress that the first seeds of the system were sown in the "war communism" period just after 1918. At this time special State Committees (Councils of People's Economy) were created to develop particular sectors of the economy. It was here that the command-economy principles of management were tested for the first time.

While the "New Economic Policy" was in operation (1924–1929) this system of economic management by the State was constantly criticized and modified but all new organizations featured the same basic centralized control systems. In the second half of the 1920s the Councils of People's Economy were reorganized into syndicates and trusts which practically monopolized certain markets and, despite their "capitalist" names, functioned as pure state bureaucratic organizations. At the start of the 1930s special "narkomats" (ministries) were created and this brought the process of creating a united, centralized management for the entire national economy to completion. Thus there was created in the Soviet Union a vertically organized hierarchical bureaucracy pervading the economic system and enabling a small group of political leaders to manage the whole economy as though it were one big factory. To simplify management as much as possible, production was excessively concentrated. Industrial development was based on the strategy of increasing production efficiency through specialization and through economies of scale. Very narrow specialization and central coordination of production units were seen as the means to increase labor productivity and the efficiency with which materials were used.

Consequently, from about 1930 the following rigid scheme was the pattern for the regulation of the economy by the State: at the top was the Central Party Committee, then a group of the central economic departments and committees, then certain executive committees and branch industrial ministries, and at the bottom, the industrial enterprises and corporations that were directly subordinate to ministries.

The main managerial functions of the ministries consisted of the formulation and direction of one-year and five-year plans and their communication to subordinate enterprises, and the coordination and implementation of the plans through the allocation of funds and resource among enterprises.

It was at the end of 1929, at the November session of the Central Party Committee, that the choice was finally made for an authoritarian system of economic management. At this session, the so-called "right wing opportunist" faction of N. Bukharin was excluded from the political leadership. Stalin and his associates gathered into their hands practically unlimited power, and the "New Economic Policy Programme", which had aimed to create a regulated market economy in the USSR, was completely rejected.

Instead, a new programme was introduced of "Accelerated Industrialization", based on collectivization of all land and farms, the complete nationalization of privately and cooperatively owned property, and the creation of heavy industries using cheap slave labor and resources expropriated from farmers.

Under these conditions, when all property belonged to the state and all democratic institutions had been eliminated along with the market economy, the state bureaucracy had complete freedom to use the nationalized means of production to further its own interests, with no interference from society, and to distribute surplus products at its own discretion.

These processes created ideal conditions for the usurpation of all political and economic power by the new ruling class, the bureaucrats.

The distinguished Soviet economist S.A. Kronrod has stressed[4] how the bureaucracy of the socialist state arises as a consequence of confiscating the means of production from their owners and then allowing a particular group within society to exercise exclusive control over them. A bureaucratic superstructure is the inevitable result. In the USSR this took the form of the state machinery of political, administrative and economic control, with its numerous officials from state and Party committees and departments, its bloated management and its bureaucratic approach.

It is informative to compare the bureaucracies of socialist states with those of countries with a market economy. The former are of course based on the philosophy of social ownership of the means of production, and the basic difference is that the socialist bureaucracy does not act as the owner, but as the monopolist of managerial function. In countries with a market economy the bureaucracy is, in fact, the owner. The rise of a socialist bureaucracy is a step in the process whereby an undemocratic superstructure develops in society, along with the abolition of democratic institutions and the deliberate elimination of the market forces that would otherwise regulate industrial production.

The Soviet experience has clearly shown how very favorable for the growth of bureaucracy are conditions where all the obligations and duties of ownership are vested in the state. The bureaucracy identifies itself as the personification of Soviet power and claims for itself the right to balance the interests of all other sectors of society.

Consequently the stratum of the bureaucracy that specialized in political and economic leadership and in top-level industrial management, became a clique which automatically developed into a special "communist" elite.

---

[4] S.A. Kronrod, Planning and functional mechanism of the economic laws of socialism, Moscow, Nauka, 1988, p. 173.

This stratum proved to be extremely well adapted for propagating itself as it developed special recruitment and training techniques – the "nomenklatura" system of accelerated promotion of candidates selected from all strata of society.

As has been stressed, the bureaucracy's main aim has been the advancement of its own interests but it certainly also desires to see a strengthening of the system of social ownership, as well as general economic development. But these other concerns only arise as a consequence of the striving for self-advancement and as there are no other mechanisms of economic control apart from the bureaucracy, it is the bureaucracy which defines the economic aims. Under present conditions the bureaucracy is not subject to market disciplines and cannot be made to accept responsibility for mistakes so the squandering of state budget money and incredibly high deficits throughout the economy are the most visible consequences of their activities. This irresponsibility can in some measure explain the very slow rate of social and economic development in the Soviet Union.

Management processes are so very centralized that it is not surprising that practically all national financial resources are firstly, concentrated in the state budget, then redistributed through budget channels to the economic units. In the 1930s about 60% of national income was distributed directly or indirectly through the budget, but by the end of the 1980s the figure was 70%.

Until 1988, enterprises had practically no opportunity to spend money without ministry permission, even money which had been placed at their disposal. Ministries themselves had to coordinate all their investment decisions with central state economic committees.

This complicated and clumsy system, with every bureaucratic body itself being controlled by another, clearly does not guarantee effective and economic use of resources. Further, as all property belongs to an abstract collective owner, namely the socialist state, there are very great opportunities to waste budget money without getting into trouble. Over the last ten years, the state has subsidized consistently unprofitable enterprises throughout the economy, to the extent of over 130 billion roubles annually. This demonstrates the almost total inability of today's bureaucratic state machine to run the national economy properly.

2. Economic Consequences of the Inefficiency of the Industrial Bureaucracy

It was in the second half of the 1970s, and in the early 1980s, that the problems arising from bad management throughout the government and the economy first became clear. At this time, the Soviet economy had practically come to the end of its "extensive" phase of development. The long-established

Soviet bureaucratic machine showed itself to be incapable of restructuring the economy so an "intensive" phase of economic development could be started. With total state ownership, a direct planning system, and bureaucratic management, it was impossible to continue to navigate onwards the ancient and rusty "ship" of the Soviet planned economy. In the first three or four years of Perestroika the government tried to intensify the economy using administrative methods, sometimes in combination with economic levers. The result was, indeed, negative; in fact, the very timid and self-contradictory steps that were taken merely pushed the country further into a major economic crisis with enormous losses in consequence.

Several economic phenomena clearly indicate the current ineffectiveness of governmental and industrial bureaucracy.

(1) increase in the number of unprofitable enterprises, and an absolute and relative increase in the amount of state and ministerial subsidy;
(2) increase in waste of resources, with consequential increase in budget deficit, rate of inflation and shortages;
(3) worsening deficit of consumer goods and other manufactures, a thriving shadow economy, and at the same time increasing stockpiles of unsalable goods;
(4) further growth of the bureaucracy and of the cost of its upkeep.

The economic machinery of the state worked in all practical terms for itself, without reference to the needs of society and of the people. In the past and still today there has been an imbalance in the rates of development of different industries. The basic industries such as shipbuilding, industrial construction and transport have always been developed more rapidly than machine building, production of consumer goods and the service sector, which have developed very slowly. From the start of the 1930s, between 60 and 80% of all financial resources were invested in such sectors as the defence industry, heavy industry, and the construction of new, very large installations.

Budget income became less and less because the increasing numbers of unprofitable enterprises and ministries demanded ever more subsidy. Practically all the profits made (75–90%) by successful enterprises were confiscated by the ministries and by the state, so there was no stimulus to cut losses, or to improve the productivity of labor by working more efficiently or intensively. Consequently the contribution of industry to budget income continued to decline. From 1965 to 1987 the proportion of budget income that was contributed by industry fell from 75 to less than 60%. In 1988–1989 the government tried to compensate for this by increasing revenue from the trade sector, mainly through increasing sales of alcoholic liquor, from cooperatives and from other non-state owned bodies. The latest new idea (i.e. at the end of 1989 and the start of 1990) has been to organize some kind of price and money reform and to expropriate money from the so-called "shadow economy entrepreneurs".

The shift towards a market economy (or, in the terminology of socialist political economy, towards "economic methods of management") has been so slow that the imbalances among industrial sectors, the failures of production and distribution, and the circulation of goods, services and money have since 1985 become even worse.

Over the last two decades the government tried to cover its mistakes and its inability to build an economic framework in which enterprises could function properly, by printing more and more money. Production of consumer goods declined, in relative though not absolute terms, year by year. This led to the demand for consumer goods greatly exceeding what was available. The policy of increasing the money supply meant that between 1973 and 1985 the amount of money in circulation tripled while the production of consumer goods only doubled. From 1986 to 1989 the discrepancy between these two indices got even worse. For example, figures from the State Committee of Statistics show that the real increase of the production of consumer goods in 1986 was about 6 billion roubles (official figure 30 billion roubles) but the total income of the population increased to 64 billion roubles. This naturally stimulated inflation, which in 1989 reached about 8% (unofficial figure 15–16%).

Deficits became more and more obvious in practically all sectors of the economy. The paradox developed that practically all types of equipment were in short supply at the same time that there were huge stocks of unused equipment and materials. For example, in 1987–1988 the total value of stocks of uninstalled equipment was 15 billion roubles, including imported equipment worth over 5 billion roubles. Enormous and totally uncontrolled sums of money were used by government, the ministries and other bureaucratic institutions to cover up losses arising from wrong managerial and strategic decisions.

Over very many years bureaucratic institutions have concealed poor economic performance by redistributing resources from profitable to unprofitable enterprises. At the state level this was done by so-called "multiscale pricing", "gratuitous subsidies", and "special budget investments". This was done through practically all the money being concentrated in the state budget and then being redistributed by such institutions as the State Planning Committee, according to the requests of ministries. In turn, the ministries used arbitrary and flexible quotas to concentrate profits in central funds, to confiscate such profits by administrative means and to redistribute financial resources among subordinate enterprises.

Under these conditions there was practically no incentive for a profitable enterprise to further increase its performance, or for unprofitable enterprises to get themselves into profit; the latter are assured of constant, and free, subsidies to cover their losses.

Thus, the administrative management system works simply and effectively. Each level in the hierarchy takes complete responsibility for the business performance of the lower levels. The state alleviates all the consequences of government mistakes through payments from the budget to the ministries. Ministries subsidize subordinate enterprises to compensate for ministerial mistakes, as all such enterprises are under the direct command of ministry bureaucrats. It is clear from history that this kind of system worked more or less efficiently during the period of quick, forced industrialization, when economic management was based on coercion. But when, from the 1960s, attempts were made to replace coercion by economic stimulus, it became more and more clear that without economic freedom, enterprises will never be interested in profit and meeting the requirements of the market.

The economic situation got particularly bad in the 1980s. The number of unprofitable enterprises increased year by year, and by 1986 14% of all enterprises were defined as unprofitable. Subsidies now accounted for 12% of national income and 17% of the state budget. [5]

Through the inefficiency of the state and ministry bureaucracies, with their irrational pricing systems, there arose a lag between prices and costs. For many industrial and consumer products, as well as for agricultural products, the costs of production, or state buying prices, were higher than the selling prices. Consequently, government subsidized prices more and more, from year to year. From 1965 to 1988 government price subsidies rose from 3.2 billion roubles to 73 billion roubles. [6]

An index of administrative inefficiency is the increasing extent of corruption in managerial staff. For example, in 1987 alone over 24,000 state and industry bureaucrats were convicted for stealing state property, bribery, and abusing their official position for personal gain. Further, 20,000 go to court each year for false accounting and misleading the state financial auditors. [7]

By the end of the 1980s the general situation in the Soviet economy had become especially dangerous.

In spite of its programme of "extraordinary measures" the government has not yet managed to stabilize the economic situation. The figures show clearly that in 1989 each successive month and quarter was worse than the preceding. In the final quarter of 1989 national income and volume of production fell by 4.5%. There was no improvement in the general situation in the early months of 1990.

Many spokesmen for the bureaucracy tried to explain these processes as being the results of strikes. In fact, in 1984 about 7 million working days were lost through strikes but the losses ascribable to bad management and poor labor discipline were 9 to 10 times more significant.

---

[5] Economic Newspaper, no. 1, 1988.
[6] Communist, no. 11, 1988, p. 70.
[7] Economic Newspaper, no. 52, 1987, p. 4.

By the end of the 1980s the pattern of investment was also giving serious cause for concern. For example in 1988, according to different government decrees, 24,000 projects were completed but work started on a further 58,000 in the same period. In 1989, 26,000 projects were completed but a further 148,000 were started. These figures show that the central government bureaucracy had practically lost control of the middle ranking bureaucrats; traditional administrative methods of regulating economic development were totally inadequate. By now it is clear that attempts to improve the general economic situation are doomed to fail if there is no radical change of the system.

The existing administrative system has unique abilities to squander immense amounts of money for no visible result. In the 1970s and early 1980s the USSR was blessed with a veritable "golden rain" of petrodollars. Over ten years income from selling oil on the world market reached astronomical figures – some 176 billion dollars. [8] But now it is clear that almost all of this rain vanished into the sand. The bureaucracy spent practically all this money in support of the existing economic regime through importing food and consumer goods. From 1971 to 1985 grain imports increased 20-fold and those of meat 5-fold.

This practice of abusing western credits continued unabated into the Perestroika period. Current estimates are that the foreign debt of the USSR now stands at 48–60 billion dollars. The most alarming aspect of the current situation is that about half of the debt arises from what have been called "Gorbachev's loans". One statistic illustrating the bad state of affairs today – the Soviet Union has uninstalled imported equipment worth 4–5 billion roubles, and this figure has not been any less for many years.

It is the clear inability of the current bureaucracy to run the economy by administrative methods and directive planning that led the Soviet Union into a grave economic and financial crisis in the second half of the 1980s. Particular harm was done by the latest attempt by government to leap forward, by creating the so-called "planning market economy" in 1989–1990.

The crisis in the Soviet economy began to cascade in 1988 when enterprises were forced to adopt "pseudo-market" forms of business activity and to try to put into effect the so-called "reconstruction and acceleration" concept. Factories were given the opportunity to increase their prices which they could easily do under the monopolistic conditions prevailing. The economy then went into a classic inflationary spiral.

During the first four months of 1990 national income decreased by 1.7% but people's income increased by 13.4%. This was even worse than what had happened from 1988 to 1990 when personal income increased by 23% (105 billion roubles). [9]

---

[8] Izvestia, 25 November 1988.
[9] Pravda, 25 May 1990.

Under these conditions of deficit and administrative over-regulation practically everything, and notably food and consumer goods, disappeared from the shelves. Inflation was, officially, 7.5% but the true figure was 12–18%.

The government tried to curb the growth of wages and introduced a special tax (the so-called Abalikin tax) on the wages funds of enterprises, intended to limit the rate of increase of these funds to 3%, but this did not succeed.

However, no effective steps were taken to control the amount of money being printed by the state bank, and pumped into the economy through enterprises which demanded from their ministries ever-increasing investments and subsidies. By the start of 1990 it was abundantly clear that the "regulation programme" of the bureaucrats had failed.

The economic situation was now practically out of control and it was clear, even to the most orthodox Party and government officials that Soviet society had to adopt a regulated market economy and to embrace the democratization of political life.

But even though they have finally accepted this, the middle-ranking bureaucrats still have illusions that it should be possible to go back to the administrative system of management. They continue to put up both passive and active resistance, ranging from spreading propaganda involving such catchphrases as "defence of the achievements of socialism", to the direct sabotage and wilful distortion of the decrees which the government has enacted in order to create a market economy. The major obstacle to reform is now the inability of professional "socialist bureaucrats" to transform their way of thinking and to begin to support the emergence of entrepreneurial activity.

3. Resistance of the Bureaucratic Machine to Radical Reform

It was clear from the first stages of economic reform (1985–1987) that the one group of bureaucrats that was more or less ready to accept the new "market" rules for the economy comprised the managers and directors of the enterprises themselves. In contrast, the central economic committees and ministries preferred to adopt a "wait and see" approach and even to block radical changes if they could.

Analysis of the results of the attempted implementation of reforms over the last few years makes it clear that the top-level officials do their very best to keep in their own hands the main levers for controlling their subordinate enterprises. Further, the attempts of government to put into effect the concept of a regulated market economy have been piecemeal and inconsistent.

As a first attempt the government did try to cut the number of bureaucrats and to abolish some middle-level institutions. The 1986 Law on State Enterprises led, within a short time, to new general schemes for the organization of industry and ministries. Following these, three Union-level ministries controlling the manufacture of machinery were abolished. A further eight Union-republic ministries were redefined as All-Union. 198 republic ministries and agencies were abolished. The staff of All-Union ministries was reduced by 62,000. Over the whole Union, the state bureaucracy was reduced by 610,000, with cost savings of 430 million roubles. [10]

Without doubt the pruning of the bureaucracy is a positive development. It must be stressed, though, that unless there are also radical changes in patterns of ownership, and the removal of enterprises from the direct control of ministries, an economic turnaround cannot be expected. Staff cuts like these will give only temporary results.

A radical reform of the economic foundations of Soviet society is very difficult because of the professionalism of the bureaucracy in political manoeuvering within such bodies as the Supreme Soviet and the People's Congresses. This is why practically all the new laws relating to establishing a market economy have not worked as expected. Laws such as those on land, ownership, and the direction of enterprises, were all adopted even though they embody very many contradictory and vague clauses. This gives the ministerial bureaucrats the perfect opportunity to put the brake on positive changes and to continue to supervise all aspects of business activity. Competition, and the economic independence of the enterprises, have still not developed properly.

One might expect that the new economic laws, which were designed to promote a regulated market economy, will merely serve to reinforce and maintain the currently existing state bureaucracy.

It must be remembered that the state bureaucracy is not just an inert mass. It includes many competent and highly qualified people. It is clear that industrial bureaucrats prefer semi-radical to radical reforms, because they do not want their style and pattern of work to change. This negative and excessively cautious attitude to reform arises from the fear of losing their jobs of status.

Thus, the big changes in the top echelons of the state bureaucracy were not accompanied by changes in the Council of Ministers of the USSR, which coordinates all Soviet economic policy. Perestroika has had only one visible result on the Council of Ministers; the number of decrees adopted increases constantly. By 1987–1989, 1,500 were being adopted per year; 4–5 per day. This is 1.5 to 2 times as many as 15–20 years ago. There was, however, no improvement in the quality and clarity of these documents, or in the means of their enforcement. The Council of Ministers spends much time deciding on local and trivial problems. This is because under the hierarchical system all such decrees must be adopted at the top level.

---

[10] Economic Newspaper, no. 5, January 1989, p. 7.

The worst feature of the post-April 1985 government decrees has been that conservative officials have been able to include in them clauses which give wide scope for interpretation and which may even prevent the decrees from being properly put into effect.

The experience of the last five years has also shown that there has, unfortunately, been no real improvement in the relationships between ministry bureaucrats and the industrial managers. This is basically because state ownership helps to preserve traditional methods of administration.

It has even been observed that as state property has been transferred to collective and joint-stock ownership, ministries have been acquiring blocks of shares in the new enterprises in attempts to gain control of new monopolistic "companies of socialist type".

As the Soviet economy moves hesitatingly along the road of radical reform, it becomes more and more clear that without radical change in patterns of ownership, and the creation of a regulated market where all types of enterprises can compete freely, it will be impossible to defeat the very powerful bureaucratic machine. The existing system of monopolies must be abolished and new and basic economic laws adopted before an effective state administration can be evolved which would be fully responsible to the Supreme Soviet of the USSR and of the republics, and to other democratic institutions.

One must stress that this process has, indeed, already begun. The logical process of development is inexorably pushing the Soviet economy towards a system based on natural economic laws. Day by day there are appearing in the USSR new enterprises such as companies, cooperatives, joint ventures, leaseholds and collectives, which are all part of a new, and strong, independent sector.

The course of events in the near future can be predicted thus, so far as the transformation of the state bureaucracy is concerned:

– Liberating enterprises from the system of state ownership will create opportunities for transforming the bureaucratic machine, and the traditional ministries will become state economic agencies promoting the development of enterprises which are partly or completely state-owned.

– Granting economic independence to the republics will speed the destruction of the hierarchical administrative subordination of the republics and territories to All-Union institutions. The republics and territories will be responsible for their own economies which they will control through their own budgets.

– Enterprises will become independent of party and ministerial institutions and will work on the basis of profit and loss.

– New economic legislation will create a basis for excluding the state bureaucracy from any control over independent businesses which will work on entrepreneurial principles. The responsibility of state agencies will be restricted to the planning of the budget.

It is the author's opinion that the progress of reform does, indeed, show that in spite of the efforts of conservative state bureaucrats to deflect it, economic and political realities are leading, step by step, towards the dismantling of the bureaucracy and the creation of a better administrative system.

Chapter XI
Intensification of the Soviet economy

V.L. Perlamutrov

Professor of Economics, Central Economic-Mathematical Institute of the USSR Academy of Sciences

1. The Essence of the Problem

The Soviet national economy has reached a turning point. It is more or less typical for the present-day world economies in general – the transition from mainly extensive to predominantly intensive economic development. The difference is only in the urgency of vital change and, naturally, in the specific ways of tackling the problem. In extensive development the consumption of resources grows faster than the output of produce. Intensive development, on the contrary, presupposes production grows faster in relation to the amount of resources utilized.

The first industrial revolution of the late 18th and early 19th century in England, followed in other countries, too (in Russia – beginning with the second half of the 19th century), initiated the replacement of man's physical labor by machines. "The first stage of industrial revolution consisted in finding machines which could replace muscle power, such as the steam engine, and also to substitute the repetitive movements of skilled hands. This was first done with the help of spinning machines and looms." [1] The productivity of human labor was the basic quality indicator of economic growth. After the dramatic qualitative transition from craftsman to factory worker machinery and technology progressed slowly, by means of partial and only occasional improvements. The weaver's son attended to the same looms used by his father and grandfather. Everything or almost everything thus depended on the skill of the worker, his experience and quickness.

The second, popularly designated "scientific and technological" revolution is conventionally dated from the first atomic explosion in 1945 (Alamogordo, NM, USA) and put an end to this state of affairs. Ever accelerating and spreading, the development of technical, production and power facilities acquired an unprecedented and completely new quality. New and even more advanced machines and mechanisms which were much more efficient and economical

---

[1] J. Bernal, World Without War, Moscow, 1960, p. 63.

were designed, manufactured and generally adopted at an extremely high pace. J. Bernal noted that "... in principle thanks to automatization, especially in connection with using electronic measuring and counting devices the production process becomes much more speedy and precise than with any machine managed by man. This can lead to a new jump in productivity which will lead to increasing of real wealth ... This means also that because of very speedy new machines it will possible to produce more and cheaper equipment and consumer goods." [2]

The age of intensive economic growth began. This country stayed too long at the start. The rapid, although largely extensive economic development of the USSR from the end of the 1920s till the beginning of the 1960s, was mainly through the use of much greater inputs of natural resources and labor. The record rate of growth of industrial production of those years began to slow down and even to fall. The appreciation that changes were imperative in the planning, financing, structuring and remuneration of labor, in the status of enterprises themselves and in the whole management system proceeded inconsistently and with difficulty. The old management methods that took shape in the 1930s and complied with extensive economic growth were incapable of adapting themselves to the newly emerging general economic situation. This situation demands economic management under which the results grow faster than the costs, when success is secured "not with number, but with skill", as Russian Field Marshal Suvorov used to say. It is evidently also to the point that extensive economic management has over the course of centuries been characteristic for Russia with its vast expanses, resources of land, water, forest, mineral deposits and labor force; to a greater extent than to most countries of West Europe. [3] Only the manifest and ever increasing shortage of coal, timber, oil and land that had previously seemed unlimited made the public aware of the fact that there could be an end to extensive development. This new altitude is evidence not so much of the shortage of resources as of the need to utilize them in a rational and civilized manner.

The capitalist economy began to adjust itself to the new situation earlier, at the end of the 1950s and the beginning of the 1960s. This process is still far from being completed. The Soviet Union, like other socialist countries, is just embarking upon the road of necessary changes. Generally speaking, it is certainly easier to follow than to pioneer. However, by virtue of essential differences in the basic principles of the system and society as a whole, and of its economy, simple adoption and borrowing of another's practices will never suffice. It means that we must summarize and assess the whole wealth of

---

[2] J. Bernal, World Without War, Moscow, 1960, p. 64.
[3] For more information about the subject, see V.O. Klyuchevsky, A Course in Russian History, Petrograd: Literary Publishing Department, Narkompros, 1918, Part 1.

experience accumulated by mankind, seek, find and try out our own ways of getting out of the present situation and advancing.

Experience, both positive and negative, of managing the socialist economy has been gained and constantly enriched in many countries. This permits theoretical generalizations and subsequent practical conclusions. This experience has, as yet, not been thoroughly studied and summed up, as a practical search for of the general regularities of development or deviations. It is only on such a scientific basis that a theory can be developed which would serve as a realistic guide for society towards the heights of effectiveness and humaneness – the milestones of a collectively organized socio-economic system.

At turning points of social development, a need always arises to return to the sources and to look over the whole evolutionary path of philosophical thought and human practice. It has always been so – in the revolutions of the 17th century in England, of the 18th in the USA and France and of 1917 in Russia. Since we are concerned with the theory and practice of socialist society, the 16th century, the beginning and the second half of the 19th century and, of course, the 20th century make the focal points.

The first theoretical monograph on a just and humanistically managed society was the Englishman Thomas More's "Utopia" [4] published in 1516. Here, a society of happy people is described for the first time ever. In this society, private property has been abolished and not only has the equality of people in respect of consumption (as in early Christian teaching) been defined, but also the production and distribution of goods, even aspects of daily life (public canteens and such like) have been socialized. Daily work for six hours is everyone's duty in Utopia. The political system is based on government by senior and experienced citizens. The author was soon promoted by the King to be Lord Chancellor of England, though not for his literary merits.

More's compatriot Robert Owen[5] was the first practitioner of socialist management. He based in his plan on the idea that "the miraculous power of machines" could provide an abundance of material benefits if private property and private accumulation of wealth were renounced. Such a concept of society would ensure real self-government of workers and would show its economic and moral superiority in practice. The great reformer strove to accomplish this in practice. Between 1820 and 1840 he set up two colonies, New Harmony in the USA and Harmony Hall in Britain, where volunteers worked on the basis of social ownership of property. Owen's enterprises did not last long, but they did demonstrate models of productive and humane labor with a force striking for the 19th century. So striking that even a sovereign – Emperor Nikolai Pavlovich arrived for a talk with the pioneer.

---

[4] Th. More, Utopia, Academia, 1935.
[5] R. Owen, Selected Papers, Moscow, 1950.

More offered the initial theoretical concept, whereas Owen showed the practical workability of an enterprise on socialist principles. There was yet another socialist experience, true, a very short-lived one – the few weeks of the Paris Commune in 1871. On account of the Civil and Franco-Prussian Wars waged at the time, political and military measures were the main actions of the Commune. Regardless of this, F. Engels, as far as it can be judged, paid attention to that Commune's decree providing that the enterprises and workshops whose owners had fled were to be leased to the workers of those establishments on behalf of the Commune. It means that they were actually turned over to the collectives as socialized property.

The scientific theory of socialist management and the practice of transition to it was developed by Marx, Engels and Lenin. From the experience of the Utopian socialists but distinct from them, they realized that a justly and collectivistically managed union of workers was feasible neither in a town nor at a factory but on a society-wide scale. The socialization of the main means of production is its natural material basis. Only a socialist revolution, with the expropriation of the means of production from the capitalists (in certain conditions it could proceed peacefully, for instance, by way of nationalization) and socializing them, can bring this about in society as a whole. The form of the economy and of society, is the same as the dominant form of property ownership.

Socialist society has two fundamental characteristics: it develops purposefully toward a deliberately defined goal – "to provide for all members of society through social production material conditions of living that are fully sufficient and improving every day together with full and free development through the use of their physical and mental abilities." [6] Its members become not only workers but co-proprietors, co-owners of the main means of production – "socialist society is one big cooperative". [7]

It is a general law of development, and like any other socio-economic law, it is effected through general environmental conditions (particular conditions of development in a given society at a given time, historical and cultural heritage, and such like) and through human individual characteristics (decision-making persons and institutions, their ability to comply with objective needs or to oppose them, talents, will, etc.). [8] Plekhanov defines the interaction between the environmental and the personal as follows: "There is a specific logic in social relations: when people are in a given set of mutual relations, they will feel, think and act in a certain way. Any political leader would struggle against this logic without success: the natural process of development will confound all his efforts. But if I know in what direction social relations will change, because of current

---

[6] K. Marx and F. Engels, Collected Works, Vol. 20, p. 294.

[7] V.I. Lenin, Collected Works, Vol. 37, p. 230; Vol. 36, pp. 161, 162, 185; Vol. 37, p. 413; Vol. 5, p. 373.

[8] G.V. Plekhanov, Selected Philosophical Papers, Vol. 11, p. 332, Moscow, 1956.

changes, then I know in what direction social psychology will change, thus I have the opportunity to influence this psychology. Changes of the 'economic conditions' whether fast or slow force on society the necessity to change its mechanisms of state. Such constant reconstruction needs the active influence of people." [9] All this taken together determines the actual development of relations.

The principal social contradictions always focus on ownership of the means of production. The main socio-economic contradiction of socialism, in my opinion, stems from the dual position of the worker as an employed hand and as a co-owner of the socialized means of production. The fundamental meaning and purpose of the revolutionary transition from private to socialized ownership of the means of production resides exactly in the resolution of this duality. Naturally, new proprietary relations produce many contradictions. They are more complicated with socialized ownership than in exploitative social structures where the parties to the contradiction are "geographically" separated, i.e., they are personified in the antagonisms of class and social groups.

The contradiction of socialism affects primarily each member of society and collective and only secondarily the classes and sections of society. While private ownership was dominant, methods of resolving socialism's difficulties were, naturally, not investigated. Resolving the basic contradiction of socialism is therefore more complicated, less "beaten and easy", to quote V. Mayakovsky. As a matter of fact, it is this contradiction that inhibits the advance of society to the stated goal of socialism and communism. Here is the cross-roads of the general, the specific and the individual. Of course, the fixed circumstances of social development can either accelerate the approach to this goal or divert from it, but they cannot change the general course of a natural and historical process. Essentially, the stagnation of the 1970s and the first half of the 1980s is the effect of real disregard of the basic contradiction. In general, the predominantly administrative methods of managing the economy are exactly those of ignoring the co-proprietary status of the worker, and hence constitute the contradiction itself. It was tacitly assumed that objective laws of development would do the required work "by themselves".

Over the years of socialist construction, it so happened that the co-proprietary function of the working people failed to acquire virtually any appreciable degree of significance or any advanced manifestation. The initial and rather primitive ways in which the socialized economy was managed were molded mainly in the first five-year plan for the purpose of the country's forced industrialization. These were for a long time visualized and understood to be immanent to socialism. The worker was cast in the role of worker only, the mere performer of a task. This was the root cause of miscalculations and errors in

[9] G.V. Plekhanov, Selected Philosophical Papers, Vol. 11, p. 332, Moscow, 1956.

management of this basic of activity. The talk was usually about "molding the attitude of the worker to be that of a master". You can only mold what has a material basis to it. Otherwise we would have to return to the 18th century, "but we are materialists after all. And the workers are materialists." [10] As a result, alienation of the worker from the means of production, which is not at all inherent in socialism, took place, which is aptly defined by the word "mismanagement". So far "the working people" have not become "the joint owner of houses, factories and tools", as F. Engels phrased it. [11] All economic reforms in the Soviet Economy, be it the New Economic Policy (NEP, 1921) or the present radical reform (1987) have two features in common. The first is that socialized ownership of the means of production dominates. Society is looking for the most efficient methods of management within exactly this kind of proprietary relations. However for decades there was no progress in these relations. The nationalization of plants, factories, land, transport and banks in the first months after the 1917 revolution, supplemented later with cooperative enterprises (collective farms, consumer unions, industrial and housing co-ops), was long thought a necessary and appropriate measure to set socialized management going. Over the past few years diverse co-operative and contract enterprises, family leaseholders and individual entrepreneurs have come into being. Considering though that these changes do not yet embrace the proprietary relations "in the vertical plane" – from the enterprise to the national economy as a whole – they cannot, in my opinion, be regarded as an actual development of proprietary relations. These are only quantitative but not qualitative changes. And this is not enough, as the practice shows, for a transition to intensive economic growth. The second feature is that economic planning and management – the leading constituent of socialized management – was and is being built as a system of physical volumes of output and distribution of produce where the tasks are measured in units, tonnes and meters. Hence the unsolvable contradiction in the management system: from "above", the tasks are given in physical units, in "kind", while in the enterprises and associations attempts are made to establish, to a lesser or greater extent, commodity, money and market relations. The national economy however is always united. In the postwar years the internationalization of national economies proceeded at a growing rate. It was either a wholly market economy or a wholly "by-the-piece" one. The internal contradictions of the NEP did not become apparent to their full extent because of the short period it had been in existence. If it had existed sufficiently long, its fate would probably have depended on whether national economic planning "in kind" would turn into planning with an emphasis on the country's money supply or perhaps the cost-accounting and self-financing trusts and syndicates

---

[10] V.I. Lenin, Collected Works, Vol. 42, p. 212.
[11] K. Marx and F. Engels, Collected Works, Vol. 18, p. 278.

would be returned to their position of 1920 when they received tasks in physical indices and their performance was evaluated by those very "subsistence" criteria. Ice and fire cannot co-exist for long. The same contradiction was characteristic of the 1965 economic reform, too. An attempt was made then to revive cost-accounting at enterprises and associations while fully preserving traditional economic planning. Although the reform was officially known as "the new system of planning and economic encouragement", the planning of the national economy remained the same. And this predetermined the general outcome. The central economic management body (the State Planning Committee) saw the shortcomings and errors in the enterprises and branch ministries, but refused to admit their existence. This factor, let alone other, less significant ones, "derailed" the reform within 2–3 years. The poet A.T. Tvardovsky defined very precisely such a situation in his "Vasily Tyorkin": "Towns are surrendered by soldiers, the generals seize them ... " The attempt to carry out another reform in 1979 had the same outcome and for the same principal reason, in spite of the numerous differences in the general and particular economic and social situation in the country in comparison with 1965. The only difference was that the fruitlessness of the changes undertaken became obvious much earlier – at the end of 1979 and the beginning of 1980. This time even the soldiers were not blamed: it was as if there had been no decisions on the reform. The ripples in the water went out and there was – silence ...

With this experience in mind, it is now possible to take a broader look at the trends and problems of the present economic reform. Radical restructuring of the proprietary relations is probably needed in the national economy as a whole, with special emphasis on changes in "the upper echelon", first and foremost in economic planning and management. Self-financing, the lease of means of production, co-op enterprises and management by contract cannot solve the problem by themselves.

2. Who is the Owner?

All presocialist societies were organized more or less similarly and uniformly. There were owners of land, water, livestock, irrigation systems, ships, machines, factories, money, banks and everything else. There were doers – slaves, craftsmen, serfs, free farmers and workers. Between the owners and the doers there was always the "spacing" of supervisors and managers. They could be slaves, shop stewards, hired foremen, engineers and other specialists employed by capitalists. Encouraged in many different ways to work better, they did so for the owners and thus for society. Society was always represented by the owners.

The socialist revolution and socialization of the means of production pursues a single purpose: to make each doer also an owner, to make all doers joint owners – co-owners. This is the basis that can make free forced labor. Everyone is involved in taking decisions and everyone bears material responsibility for the end result: the better we act, the better we live. And on the contrary: errors in decision-making or poor performance leads to an inefficient economy, and hence, needs remain unsatisfied, with incomplete social protection for the people. Miracles do not happen in the economy. There is not and cannot be stimulation there in the way that was practiced for centuries and millennia: the owner, naturally, neither encourages nor punishes himself and directly and indirectly he lives upon the aggregate labor. The socialist system of production is more effective and humane than any other. The matter is in how to put it into effect and how to make it work.

K. Marx called the society of the future "the association of free producers" [12], "the association of free and equal producers engaged in social labor under a general and efficient plan" [13]. At the end of the 19th century F. Engels made a more specific assertion. In his article "On Housing" [14] he wrote that under socialism the "associations" (the collectives of doers) would lease from society for payment the means of production required for work such as plants, factories, houses, workshops. It would be unlikely that the use of these facilities would be granted, at least in the transition period, without payment to cover expenses. At the same time eliminating private ownership of land would not entail abolition of rent but use of the land would be paid for, to the benefit of society. Thus the taking possession of all means of production does not exclude the operation of hire and rental procedures. Rent does not at all mean gradual redemption of property from the owner. It is the compensatory placing of facilities at someone's disposal on mutually acceptable terms specified by an appropriate agreement. The lessor (in this case, the whole society) unquestionably remains the owner, provided of course that it takes no other decision.

Later socialist thought seeks and finds more specific, understandable and practically feasible forms where these general laws of future society manifest themselves. From its very start as Russian social democracy the revolutionary movement in Russia was extremely active in following, studying, participating in, learning and evaluating developments in the world socialist movement. "The cooperative apparatus is the apparatus for supply, based not on the private initiative of capitalists, but on the mass involvement of workers, and Mr. Kautsky, when he was a marxist, was right, when he said that socialist society is one big cooperative". [15]

---

[12] K. Marx and F. Engels, Collected Works, Vol. 4, p. 447.
[13] K. Marx and F. Engels, Collected Works, Vol. 4, p. 57.
[14] K. Marx and F. Engels, Collected Works, Vol. 4, p. 278.
[15] V.I. Lenin, Collected Works, Vol. 37, p. 230.

Two matters were in the limelight in the first months after the socialist revolution in Russia: the spreading of Soviet power to the whole country and the conclusion of the Brest–Litovsk Peace Treaty with Germany, the Austro-Hungarian Empire and Turkey – a burdensome, humiliating and "shallow" one, as they said at the time, but necessary. Peace was concluded in March 1918, and so soon as April V.I. Lenin wrote "The Next Tasks of Soviet Power". In it, as in a number of other post-October works and speeches (including "On Cooperation", one of the last three works dictated in early 1923), Lenin invariably propounds the thought: the basic task of Soviet power is getting the general cooperation of the population for the purpose of turning all citizens into members of one national or, more exactly, nationwide cooperative: the system of civilized cooperators under socialized ownership of the means of production is socialism; socialism is one large cooperative, i.e., an association of people where everyone – the employers, owners and workers are doers.

It was in this that the general idea was expressed and the main road to socialism was conceived, with the particular route for such a backward country as Russia to follow. Subsequent interpreters narrowed down the meaning to one particular, though significant aspect – cooperation in agriculture. Of course the actual process of social development always shows how many contradictory principles, situations and circumstances have emerged to slow it and even to divert it from the main road (Soviet history abounds in them), but that is why a general law is a law, because it manifests and asserts itself and comes true in one way or another. Otherwise social development would always be a chaotic conglomeration of chance occurrences.

The fact that in recent years the long domination of the "administrative-command" system over economic methods has been much and variedly talked and written about is definitely a step in the direction of the truth. If society manages its economy not as the most democratic of all economic systems, not as the union of doers and co-owners, not as a united civilized nationwide formation of cooperators, there cannot be "factual mastering of all implements of labor on the part of the working people", [16] but only formal socialization, which gives scope to voluntarism, red tape and "bureaucratic departmental administration" [17] (a definition given to the economic mechanism of the "war communism" period at the 12th Party congress), or to bureaucratic state-planning administration, as it should be clarified now. Furthermore, it enables the omnipotent and arbitrary rule of the bureaucratic state machinery to function outside any kind of open control by the people and by democratic institutions.

---

[16] K. Marx and F. Engels, Collected Works, Vol. 18, p. 278.

[17] The CPSU in the Resolutions and Decisions of the Congresses and the Plenums of the Central Committee. Moscow, Politizdat, 1954, Part 1, p. 692 (in Russian).

A rather strange economic system thus emerged, with nearly a hundred and thirty million doers and a few tens of millions of supervisors and managers wielding different degrees of authority and power. And virtually no proprietors – in society where everyone is supposed to take part in making decisions and be responsible for their results as collective owners. It means that not a single person, by virtue of his or her objective status in society, is concerned about taking constant and daily care of the efficient spending of each rouble of the public wealth, about preserving and further augmenting social property. It is exactly such a situation which were characterized in the classics of socialist thought as alienation of the working man from the means of production and were attributed exclusively to presocialist societies. The practical result was that alienation extended to the socialist economic system, too, and in almost its worst form: in previous societies the owners did function and did pull efficiency upwards in accordance with their own particular and indispensable function.

The feudal landlords put the windmill and the mechanical clock to work, the capitalists the steam engine and the mechanical loom, and later on many other useful machines ... But the society of co-owners of the means of production by virtue of its nature actually became a society of hired doers working merely for their wage. Naturally, the managers throughout, from the team, factory and plant to the State Planning Committee, substituted for the absentee owner. They are those whom the Soviet press has sharply and spitefully exposed as bureaucrats in recent years, identifying them as the origin of all social troubles. After the first and understandable confusion the bureaucracy is gaining experience of working in the new conditions: it argues with the press and continues its routine work in silence whenever possible. And it will continue to do both until it feels the strong mind, will and hand of the owner of the socialized means of production.

3. Democratic Choice of the Paths of Social Development

The idea that the governing body automatically, as a matter of fact, mirrored the interests of those sections of people it represents was long current with Soviet social scientists: the bourgeois state represented the interests of the bourgeoisie, a proletarian state those of the proletariat, the managerial board of a Soviet enterprise those of the collective, etc. In the absence of real co-ownership by the doers, economic planning and management devolves entirely on specialized state bodies. Specialization and professionalism are certainly needed to arrive at correct decisions though this generates contradictions too: "Society cannot function without certain specialized workers. These create a new division of labor within society. Though they have been entrusted with their specialized tasks by their superiors they become independent from them." [18] In the absence of constant public control of their activity, and of any realistic possibility of replacing them these authorized individuals inevitably place their own interests above those of the society that they recruited to serve.

Even if we assumed for the moment that the governing body did not pursue its own interests, it does not change the matter – the weakness of democratic institutions leads to decisions about the development of society being taken not on the basis of the true purpose of society, but in terms of the fulfilment of particular targets, as visualized in the minds of functionaries. It is not by chance that capitalism, for example, developed over its long history a rather involved, but smooth system of bourgeois democracy ensuring power to the bourgeoisie. By the way, the system basically features rather broad political democracy (rule by elected authority) which in no way extends to the economy. The omnipotence of the owners of capital is complete: they govern. In this situation, political power is also in the hands of the owners. It is invariably in their hands even when the workers, socialist blocs and parties and not the bourgeoisie win a majority in parliamentary elections.

Democratization is not a form of socio-economic relations that is exclusive to socialism. Democratic management of the socialized economy can become a permanent, firm and stable basis for the democratic development of a socialist society without the diverse and mutual influence of politics and economy. It means first of all that major economic, regional and factory decisions elaborated by specialists (no one else can do it) are considered and adopted by the doers, members of collectives themselves either directly at meetings and by referenda or through their representatives at conferences and congresses. Society based on real socialized ownership does not and cannot have any other road. Naturally, the proposed decisions must be presented as a choice of several versions – otherwise the opinion of specialists alone is heeded to. Aware of these dangers, K. Marx [19] in fact criticized M. Bakunin who regarded democratic elections of the authorities as the main device of the socialist restructuring of society.

In the years of Perestroika the transition to economic management methods is intimately associated with the democratization of public life. There are however many obstacles on this path which can hardly be surmounted until the traditional forms and methods of economic management are radically changed. Without this, the economic methods themselves and their actual basis – the development of market, commodity and money relations, will remain on the level of good resolutions or vague notions. This can be put more definitely: commodity and money relations are but one of the elements of the democratic mechanism, a secondary and derived one to boot. There is no room, as a rule,

---

[18] K. Marx and F. Engels, Collected Works, Vol. 37, p. 416.
[19] K. Marx and F. Engels, Collected Works, Vol. 18, p. 616.

for realistic commodity and money relations where democratic command of "the whole socialized capital" is lacking so far as public property is concerned.

In the present stage of socialism it is therefore highly important to look for new methods of management that would not be limited to the previous experience and practices. The criteria that innovations in management must meet relate to their effects upon the doers and their collectives for the purpose of rationalizing the whole course of economic affairs, i.e. improvements must be sought in the present state of affairs towards greater effectiveness and more complete satisfaction of the public needs.

The democratic procedure of choosing the country's socio-economic development route serves as a reference point. Socialist democracy is not merely democratic voting and formation of the bodies of power. On the economic scene, it implies free, open and creative discussion of the previous stage of social development, an analysis of its shortcomings and unsettled problems and identification of any undemocratic tendencies, all of which will help it to enter the next stage. The most effective directions for further development are chosen in a responsible way by the members of society as owners of the socialized economy. The discussion and decision-making should concern primarily fundamental questions of long-range strategy such as relationship between consumption and saving in the national income, the share of expenditure on social needs, science, defence, state security, investment, and banking policies, etc. Decisions are needed on the implementation or non-implementation of projects and programmes of major national or regional economic, ecological and social significance; on the singling out of social benefits (at present, included among these are housing, health services and education) which it would be sensible to distribute not purely according to the labor contribution of the doers; on the taxing of the earnings of enterprises and citizens, and such like. It is a matter of principle to adhere to the democratic procedure of establishing general rules for economic management and introducing corresponding legislation.

The idea that the above functions could be effectively performed by specialized state bodies has long been current in Soviet economics and practice. It was no accident that from the world's first ever long-term plan of national economic development (GOELRO) all plans involved offering only one option for each scheme, so there was no discussion, only approval. The working people were thus cut off from participating in economic management, and, as mentioned, the interest and notions of managerial groups replaced for the goals of society.

The transition to democratic practices in the economy will allow orientation of planning towards securing balance and harmony among the groups and strata of society, and considering the actual needs and possibilities. As for the behavioral motivation of doers, this is effective under socialized ownership provided the production collective and the doer identify themselves with society as a whole, i.e. with the status of owner not being confined to the manager. The doer as co-owner of the socialized means of production is a participant in the economic process who is engaged in fulfilling not tasks from "above", but in participating in both the definition of objectives and in their implementation. This is where the working people and the collectives play an adequate part in the socialist process. Such a role is but another expression of the doers' co-proprietary status in society.

The long-existing and manifest lack of appreciation of the importance of motivation in management led, and continues to lead, to the plan-governed socialist economy being interpreted chiefly and even exclusively as a system of centralized and physically voluminous planned tasks "apportioned" from top to bottom across industries, enterprises and regions. In spite of the peremptorily proclaimed urgency of the planned apportionments they were ignored in many cases. There has not been a single five-year plan whose target figures were met. The same is largely true even of the eighth five-year plan which was relatively successful. Commodity and money methods of management, and market methods in general are alien to this situation of subsistence "apportionments". The theoretical views of socialism which had long ignored the market methods under socialized ownership followed, one should think, the same direction. The consequences were such that neither in drawing up the plan nor in its fulfilment was anyone concerned about the correspondence and balance of commodity and money circulations in the economy (though exceptions were made because of obvious necessities in the consumer sector). Without such a balance, the market levers of management are denied freedom.

Incidentally, opinions are rather often voiced in the press that the monopoly position of the sectoral ministries and large manufacturers is against the interests of consumers, leads to rising prices, lower product quality and other negative phenomena. The transition to the market forms of management is seen as an escape from this. In such an approach, I think, everything has been turned upside-down. It is not a matter of whether many or few manufacturers make one product or another: it is a secondary factor. Monopoly is first and foremost the result of the solvent demand of consumers exceeding the supply of commodity from the producers. If the demand is steadily ahead of the supply, it is a sellers market regardless of whether there are many or few of them. The situation in itself predetermines the deficiency of the economic process. No matter how many producers, they are not in such a case forced to compete for "favor" from consumers. On the contrary, the buyers line up for their services. The producers' competition for consumers, i.e. the latter's supremacy in a buyers' market, is possible only when there are balanced and corresponding flows of commodities and money in the economy. Hence the paramount significance attached to the planning, structuring and functioning of the country's monetary system. The changes must start with the reorganization of national economic planning.

A new pattern of planning is needed that would unconditionally ensure both the definition and the implementation of consistent stages in achieving the socio-economic objectives of society. It must develop the democratization of intensive economic management and envisage various alternative plans for socio-economic development with an optimum one being chosen by democratic procedures with due regard for assurances of the citizens' free will. A particular alternative plan is not obligatory for all enterprises, but it serves primarily as a basis for the government's economic policy as well as one of the leading landmarks for the independently managed economic units whose functioning must also be based on the same democratic principles as the national economy.

The existing banking and credit system opposes the switching of the economy to steady intensive growth, because it is based on forced "pumping" of money into circulation. In fact, this makes industries, enterprises and regions uninterested in the efficient utilization of the productive, raw material and labor potential and orientates them towards ever higher consumption of resources. The money reserves for planned turnover are largely provided by excessive increases of credit. This gives rise to credit inflation which speeds up the growth of current and capital expenditure on the production of commodities, creates unnecessary material reserves in the economy and increases the number of unfinished construction projects.

The more resolutely the decades-long stereotypes of extensive management will be rejected, the more effectively the financial and credit reform will proceed. And the other way round, their survival can seriously undermine the reorganization of economic management. It concerns, in the first place, the vocabulary of the planners', managers' and financiers' thinking. Beginning with the '30s, output expressed in units, tonnes and meters dominated the national economic plans and the appraisal of the performance of enterprises and industries. Money, finance and credit acted as auxiliary, secondary economic instruments in those conditions. Generations of Soviet managers were brought up to believe these notions self-evident and indisputable.

The situation demands radical changes. The philosophy of reform is that money must be given back its natural function of serving as a general equivalent to material values, just as Marx defined it in "Das Kapital" [20]. For the purpose of intensifying the economy, it is money that is the objective witness of the social value and usefulness of products. Payment, or refusal to pay, for delivery or service, profitability or unprofitability of an enterprise are statements of this value. The manager now has to combine the functions of an economist and a financier.

The "stereotype of departmentalism" is particularly dangerous: it maintains that the restructuring of planning is the business of the State Planning

---

[20] K. Marx and F. Engels, Collected Works, Vol. 23, p. 120.

Committee, that of material and technical supply of the State Committee for Material and Technical Supplies, that of crediting of the banking system, etc. Practice has shown that all more or less major national economic problems are interdepartmental. Intensification of planning will succeed if it embraces the material, labor and monetary resources altogether. The State Committee for Material and Technical Supplies alone cannot cope with the transition from supply by quotas and warrants to wholesale trade in the means of production without activating the financial, credit and price levers of management. The financial and credit reform is not a departmental one, but a national economic reform.

The "stereotypes of science" on finance under socialism also conceal great dangers. The theoretical propositions on which it is based and which are taught to students were formed mainly in the years of developing "subsistence" economic planning and management. The fiscal relations in the economy were assigned a purely subordinate part. The principal task was fully to finance the expenditure envisaged by the plan regardless of any assurance of its profitability. By tradition, the science of finance proceeds from the priority of material resources over monetary ones. For instance, it deals, in the main, with material current assets and very little with assets in settlements and financial obligations. It is unconcerned about the fact that the amount of overdue payment in the economy, particularly for bank credits, is growing faster than other economic indices. This is nothing else than a display of irresponsible use of all resources by enterprises and industries – material, labor and monetary. Furthermore, this science like other economic disciplines studies the fixed assets of enterprises, industries and the economy as a whole and the fixed assets themselves in terms of how they should follow from the science itself. As a matter of fact, money turnover is almost entirely a "blank space" for it. This turnover though is to become one of the major objectives of planned management.

The use of "socially owned capital" degenerated for a long time. This was equally true of the fixed capital (where the returns on assets decreased) and the current one (the turnover of the circulating assets slowed down). Each rouble invested in the economy yields ever smaller returns. Any measures for financial improvement will have but temporary and unstable results until an end is put to this tendency.

The average service life of equipment in the industry has been systematically extended and has reached 28 years. In the rest of the world, new generations of machinery and technology appear every 7 to 9 years on average. According to the Promstroibank of the USSR, the commissioning of the projects already begun will take about 6 years and in some industries even 9 years. Keeping the situation unchanged means that we lag by a generation of machinery and technology.

In order to avoid this we must narrow the already existing gap and reliably balance the material and fiscal resources in the national economic plan. The

State Planning Committee of the USSR is obliged to make up the country's five-year fiscal balance. This plan is however not a government task for the economy to be expressed in terms of volume and turnover rate of the monetary resources, but merely "a miscalculation" of already earmarked expenses with a monetary expression – on capital investments, growth of current assets and other needs. With such a procedure of planning, money cannot act to limit expenditure, but, on the contrary, it passively follows the process. It does not compel planning bodies, industries and enterprises "to cut the coat of monetary resources according to the cloth of economic expenditure", but boosts increased and excessive material and labor costs. A new method of economic planning is therefore required. The plan for the 13th five-year plan period must certainly be developed as a balanced commodity–money one. Such a plan will provide an impetus to the cost-accounting, self-repayment and self-financing of enterprises and associations.

Theoretically, the following aspect can hardly be questioned. If socialist enterprises act as commodity producers, the national economic plan must also turn from one based on kind (where the basic indices are expressed in units, tonnes and meters) into a commodity–money one where the amount and the range of commodities are countered by the monetary resources in the hands of consumers – the industries, enterprises and population, or where the commodity supply corresponds to the effective demand. The "compatibility" of planned and cost-accounting levers of management is thus secured, i.e., the integrity of the economic management system is assured. Otherwise the plan and cost-accounting will "reject" each other, which was, by the way, one of the reasons for the failure of the 1965 economic reform. It would be erroneous to reduce the reform of planning to cutting back the number of material balances worked out by the State Planning Committee, or to transferring some of them to the planning bodies of the State Committee for Material and Technical Supplies. These would be changes only within the existing planning method. The dilemma is now: either a plan in kind and curtailed, formal cost-accounting or a commodity–money plan and full cost-accounting and self-financing of enterprises. Intermediate versions are practically non-existent.

A balanced plan is a fiscal plan in kind where each side is "equal", meaning that included in it can be those tasks for construction, commodity production, transportation and consumption that fully correspond to the amount and the turnover rate of the country's assets. The general provision is that the two principal "gears" of the planning mechanism revolve in complete and reliable "coupling". True, we no longer have planners capable of operating with such instruments. The first five-year plan alone was drawn up as a financial one in kind. But this cannot be an obstacle to the improvement of planning. Practical experience in economic planning was lacking until 1920 when the GOELRO plan was worked out. It was gained in the process itself. In his article "On a

Single Economic Plan" V.I. Lenin called the GOELRO plan a scientific study, a genuine scientific plan.

As one of the major tasks, the new economic planning procedure should probably include in the first place not only the distribution rates of earnings; profits, wages, bonuses, etc., but their basis; the rates of return (self-repayment, turnover) on national economic expenditure: depreciation, removal and renewal of fixed assets, their profitability and turnover of current assets. The planned settlements could thus link the earnings and the spending of the material and monetary resources in circulation and reproduction.

For practical purposes, all the details of every engineering and technological project (design, quality, price) and of renovation projects will not be entered in the plan. Only those that are really economical and will repay the investment in due time (close to 7–9 years) will be supported. The economic rates are to cross out ineffective options of construction, technology and marketing from the package of as yet unapproved proposals. In such a rigidly framed plan it will no longer be possible to put up with obsolete machinery, shortages of production equipment, the overconsumption of resources and "dead" stock at warehouses.

Extensive use of primarily economic methods instead of administrative ones in management is effective with precisely balanced material and monetary resources both in the national economic plan and in the course of its fulfilment. The economic levers such as prices, interest rates, profits, fines, wages and bonuses have money as their foundation and work effectively for the economy. Stringent economic methods of managing cost-accounting collectives and a weak financial and credit system are two incompatible things.

When the associations and enterprises become self-financing, it would be advisable to restrict wholesale payments over the course of a few years and apply appropriate tidying of their book-keeping. The restriction of the amount of money in circulation will force managers to get rid of surplus and old stocks of raw materials, finished and semi-finished products. Self-financing will have the same effect provided it is strictly observed. The incentives for speeding up the turnover of current assets and for efficient management will become a reality. Cleaning up the balances will lay a realistic foundation for the assessment of the actual situation in the economy.

The "compatibility" of the plan and the market and even the conditions for the formation and development of the latter do not completely exhaust the "market" subject. We must also make it clear which type of market relations is in agreement and which is not with the nature of socialist society, considering that in the past few years (or even months?) our social science has recognized the compatibility of market and socialism. As a theoretical and practical issue, this problem has come to the notice of European Social Democracy much earlier, particularly in Sweden, Austria, Finland and Norway. A highly effective but rather inhumane market economy functions in many capitalist countries featuring heavy exploitation of the countries producing primary commodities, labor-intensive production, a tremendous gap between the incomes of workers and employers, constant unemployment, poor social security for hired labor and, particularly of the youth. It is perfectly obvious that such a path does not suit us.

Neither does another path suit us, virtually hypothetical for the second half of the 20th century and long replaced in the course of economic development itself: numerous very small producers crowding the markets in the hope of selling the fruit of their labor at some profit. It would not have been worth mentioning were it not for one thing. Our present-day social and political journalism when analyzing economic subjects presents the market economy to the reader in this, or almost in this manner. The reality is far from this. After the publication of the book by the English economist J. Maynard Keynes[21] and the implementation of his recommendations by US President Franklin D. Roosevelt at the beginning of the 1930s (New Deal) [22] with increased interference of the state in business matters, this kind of market actually disappeared in the developed countries.

The issue is not that the greater independence of enterprises entails corresponding slackening of centralized management, as sometimes believed. There is more to economics than just sums. In general, an industrial economy under the scientific and technological revolution is unthinkable without a certain degree of centralization. The time is long gone when President Roosevelt's New Deal bills were repeatedly rejected by the US Congress as contradicting human rights. An economy based on socialized ownership of the means of production is also inconceivable without centralism in management to an even greater degree. It all depends on the methods of practicing centralism in our economy and in our time. Instead of "pressure of strength" they acquire a new quality – "gentle management by the conductor".

President Roosevelt disregarded the stereotypes of bourgeois thinking and took capitalist society out of the severest ever economic crisis whose social consequences could have become unpredictable. He was aware of this, and admitted in talks with his associates that if he were a bad president, he would be the last one ...

American economist John K. Galbraith offers one of the many examples of changes in capitalist management. [23] In the summer of 1903, the Ford Motor Company went into business with a capital of 28,500 dollars and 125-strong work force. By the mid-1960s, the company employed 317,000 people and its capital had reached 6 billion dollars. Instead of tens and hundreds, the annual output

---

[21] J.M. Keynes, The General Theory of Employment, Interest and Money. London, Macmillan, 1936.

[22] See N.N. Yakoblev, FDR: A person and politician, In: N.N. Yakoblev, Selected Works, Moscow, Mezhdunarodniye Otnosheniya, 1988 (in Russian).

[23] J. Galbraith, New Industrial Society, Moscow, Progress, 1969, p. 53.

of vehicles numbered millions. The growth of output itself, let alone everything else (relations with suppliers, competitors, etc.) was no longer compatible with an unknown and unorganized market.

Today, this conclusion is true to a much greater extent: it concerns the planning principle not only on a national, but also on an international scale. John Bernal advanced the idea of "planning the whole world". [24]

It cannot be said today that mankind has realized the necessity to implement this idea. Even so, and strange as it may seem, practical steps that could and will probably take us to this road have been made. Transnational monopolies are already a reality and so too are supernational organs of the world economy such as the International Monetary Fund, the International Bank for Reconstruction and Development, the General Agreement on Tariffs and Trade, the Council for Mutual Economic Assistance and the European Community with its own international currency unit – ECU.

The market differs strikingly from what it was in the 19th century generally, or till 1932 in the USA (beginning of the New Deal), or till 1922–1927 in the USSR. Mutually coordinated and controlled under a plan are not only the operations of companies and national economies, but also the leading economic processes and tendencies in the world economy. Essentially, it is the planning of markets and the "marketizing" of planning. This of course does not make it free of difficult and challenging problems, ups and downs as well as monetary and financial crises, and other upheavals. The mutual debts of countries, continents, population strata, companies and banks are so confused that a strong jolt in one-two-three links is enough to induce a chain reaction of bankruptcies ... This subject is however not pressing yet.

What is left for us as acceptable is the building of a humane market economy. In principle, it is the only one adequate to socialism. Generally speaking, the matter is that the requirements of effectiveness and humanity in economic relations are contradictory. The development of our economy in the 1930s and 1940s was quite effective with record-high growth rates but unsatisfactory in human terms. The economic development of the 1970s and the 1980s could be regarded as relatively humane, but clearly ineffective: little was demanded of the doer, and he worked ever less effectively. In the present situation, an optimum relationship must be identified between effectiveness and humanity, the social and economic protection of man and society. In the socialist countries, this problem has not arisen yet as one requiring a real solution, though it did arise immediately after recognition of the market character of the economy based on socialist ownership. The practices of Western socialists are certainly important and valuable as subjects of study themselves, with all their positive and negative aspects, but not as a recipe for adoption. It is perhaps time to

---

[24] J. Bernal, World Without War. Moscow, 1960, pp. 51–54.

move on from general talk about market economy and from an abstract trust
in its miraculousness to the search for a definite model with its two poles of
effectiveness and humanity.

Perestroika set a course for democratization of Soviet society. Regardless
of the contradictory diversity of relations in all public spheres, democratization
in the management of the socialized economy is a key issue which ultimately
brings about and determines realistic, stable and effective democratic institutions throughout society. Democracy in political, social, national and other
relations will not be able to consolidate and to consistently manifest itself, in
general, without radical changes in this area. We should not forget that the
democratization of our society began in 1956 after the 20th Congress of the
CPSU clashed, first and foremost, with the old methods of management which
no one had even attempted to change. The result was not long in coming: in
a few years universal roll-back to the old methods of bureaucratic command
and management took place in all spheres of public life: managers who are not
owners can work only as they are accustomed to. Between the mid-1950s and
mid-1980s, the changes in the management mechanism included replacement
of branch ministries by national economic councils and their re-institution,
division of the State Planning Committee into two bodies – the State Planning
Committee (current planning) and the State Economic Council (long-range
planning) and their re-emergence, the reform of 1965 and an attempt at a reform
in 1979 which referred only to the status of enterprises in the economy, but
did not change their economic infrastructure. These and other measures had
nothing in common with transition to democratic methods of management.

We must also heed a historical lesson – the democratic institutions of the
emerging bourgeois system were brought to life by the following circumstances:
inviolability of capital and the person of the merchant and of the entrepreneur,
inviolability of "capital" (banks) and of the worker's person. Capitalism came
into being and was consolidated first of all where the above conditions ruled
earliest (the Italian city states of the late Middle Ages), and then gradually
developed in the west of Europe. In any case, bourgeois democratic transformations started with economic necessity, whereas in the countries of the East the
sultans, caliphs, tsars, pashas, khans and other rulers could not even imagine, let
alone permit, anyone's personal inviolability. Economic development came to a
standstill and decayed there. Civilization, though, emerged in the East.

The revolutionary centralization of the main means of production has been
both an actual experience and a major socio-economic incentive and motivation
for generations of Soviet people (in the years of the formation of Soviet
power, industrialization, the Second World War, the postwar rehabilitation of
the ruined economy). The contradiction of "non-ownership" in the socialized
economy however was becoming ever more manifest and acute with each
passing generation. It was due to the growing "tension" within this contradiction

that society eventually crept into the critical situation of stagnation. The fact of nationalization itself, natural for the younger generations, was no longer enough for proper labor motivation and activity of the workers. It so happened that private ownership had long been banished, but the "aggregate owner" failed to perform its natural function or did it poorly and sporadically. As a consequence, we had numerous and steadily recurring cases of mismanagement, inefficiency, irresponsibility and impatience for all kinds of improvement in the economy and in personal well-being, from society as a whole regardless of individual contribution. The stability and the mass character of this phenomenon demonstrates the obsolete nature of the existing forms of ownership of the means of production.

4. Workers as Co-owners of Enterprises

As conceived by the new economic reform, the status of enterprises in the Soviet economy is to change radically. The Law on State Enterprise (association) came into effect in 1989, and was preceded by the law on cooperation in the USSR and the Law on Individual Enterprise. After the 1923 decree on trusts, the legal rules for enterprises were set mainly by executive bodies of power. The rules were not as definite and unambiguous as a law, they had essentially no backing and were frequently changed. Roughly estimating, the banking rules for enterprises alone surpass, in bulk, Leo Tolstoy's "War and Peace".

The new law envisages self-financing. If strictly observed, it will form the material basis of the enterprise's liability for its management results. The enterprise will no longer have a kind minister with a big moneybag behind its back. Kind because he himself has wheedled the money out of the Ministry of Finance or a bank. From now on, the money for current expenses and development will have to be earned by the enterprise itself. Naturally, this process takes time. Some will have to replace equipment, to improve the range and quality of products and to find reliable contracting parties, while others will have to merge with larger enterprises or to close down. Some of the prices for the produce will be set by an agreement between the supplier and the buyer. Cooperation of enterprises with contracting parties in other countries is becoming a reality. Enterprises are being set up in the USSR with the participation of foreign-owned capital. Enterprises have been granted the right of independent operation in external markets.

The enterprises are turning into self-managed units. By law, the management belongs to the general meeting of workers (or to the conference of their representatives in the case of large enterprises). The elected council of a work collective is responsible for day-to-day matters, as if it were a kind of parliament. It is a natural change – if the responsibility for performance is transferred to the collective of workers, it makes decisions on economic and social matters itself. No one else can know better and pass a better decision. The management becomes an executive organ and is fully accountable to the collective.

Soviet enterprises are now allowed to operate in foreign markets on their own and to have joint ventures with foreign partners. World practice shows that the incentives for achieving higher standards of machinery and technology are much stronger with management that is not limited to the national economies. Even securing the vanguard position within the national boundaries is only half of the matter. Admittance to the world market means capturing the top positions in scientific and technological progress. You cannot learn to swim without wetting your feet. So far we have only slightly more than a hundred joint enterprises and about a hundred of our plants have had experience of operating in the external market.

Broad and systematic participation in the world economic division of labor is the "higher" school of scientific and technological progress. Successes on the world market alone can make the rouble convertible into other currencies. So far we have been trying to do the opposite – to earn other people's money by exporting raw materials and energy resources. This path does not lead to the ultimate goal. It can be achieved only through high and ever accelerating rates of improving the material and technical basis of production, product quality, higher professionalism of the workers and their involvement in the affairs at the factory and in the country at large.

Self-management will have no realistic and tangible effect until actual self-financing is ensured, until the channels of administrative and directionary "injections" and withdrawal of monetary resources are blocked. For this purpose, at least the following changes are needed in the present economic situation of enterprises and associations.

First, a radical reform of the credit and banking system is required. It has been launched, but so far it is either skidding or moving somewhat chaotically and often, most regrettably, in the wrong direction. As in previous years, the growth of credit to enterprises and industries steadily surpasses that of output. The economy is being systematically oversaturated with large sums of money, which, let alone other factors, predetermines shortages of virtually all kinds of goods and defies the employment of economic (instead of administrative) methods of management and distribution. Wholesale trade in the means of production, an indispensable attribute of the market economy, remains among the innovations denied application for as much as a quarter of a century. The demand greatly outdistances the supply of goods and services and thus frustrates interest in scientific and technological progress. In the final analysis, this is all because of the subordinate position of banks in relation to the state planning institutions with their tasks in kind (units, tonnes, meters). It is also alarming that the rather privileged grants of credits to enterprises (write-off of debts, low interest rates on loans, and such like) are actually flaws in financing practice, because the shortage of money as a result of poor performance is often compensated automatically, in fact, with bank credits (a new credit can be contracted even without settling the previous one on schedule).

Furthermore, numerous new banks keep rapidly appearing (specialized, cooperative, joint-stock) and each strives to secure absolute independence – either because of its youth or its inexperience. Even the USSR Academy of Sciences is going to set up its own bank, although nowhere in the world has anyone ever given credits for fundamental research. Attempts to discover the laws of thermodynamics and the theory of relativity have never been made on bank loans. Besides, who ever founded a bank without having one's own receipts (the Academy exists mainly at the expense of the state budget)?

The role of the State Bank as the country's central bank is patently belittled. Just like the central bank of any industrially developed country, it alone can and must keep hold of and control the whole money turnover. If the State Bank fails to perform its function as a monetary center or does it poorly and diffidently in this situation, all will end up in financial chaos. In order to escape such developments the State Bank should be subordinated, like in most countries, not to the government, but to the parliament – the Supreme Soviet of the USSR, and should make its issue and credit regulations obligatory for every bank in the country. The new Regulations and Rules of the USSR State Bank (1988) however subordinate it to the government. Paradoxical as it may be, the present Soviet banking system is less centralized than that in the capitalist world. From the viewpoint of production efficiency, the cost of this practice is high and unjustified.

Although to a lesser degree, the above also applies the Ministry of Finance. The state budget approved by the Supreme Soviet must be under its constant control. One should suppose that the present schedule of the country's supreme body of power will allow time to exert it. The government has no right to spend a single kopeck above the law-approved sums and designations. Its field of competence must be limited to routine spending alone. Intensive management of the kind new to us is impossible without it.

Second, the transfer of public productive assets on lease to enterprises under a contractual agreement with a public authority and not on an administrative basis with unilateral ("from top to bottom") liability. [25] The agreement stipulates the mutual obligations: the amount, technical condition, capacity, etc., of the transferred productive assets on the one hand and the rent, procedure

---

[25] See also V.L. Perlamutrov, What is Cost-Accounting to be Like? "Trud" newspaper, Feb. 13, 1985; On the Socialist Intensive-Type Management Mechanism, In: Ekonomika Matematicheskie Metody, No. 5, 1985; Cost-Accounting in the Past and in the Future, In: Znaniye – Sila, No. 1, 1987 (in Russian).

for payment, etc., on the other. The minimum return from the assets which the collective guarantees with its labor and earnings is thus specified together with its managerial effectiveness.

The Law on Lease (1989), however, does not settle the whole question. Incidentally, such a major problem as the relations between the leaseholder and the lessor is yet to be dealt with. If the latter is represented by the present branch ministries which are merely administrative establishments, real market relations will not form. The ministries lack their "own property" with which they could be held liable for their performance. If they are allotted such property, they will become entirely different organizations. Besides, new legal rules on their activity are needed, too. As to land, water and mineral deposits, it would be natural for the local bodies of Soviet power – the Soviets of People's Deputies – to act as the lessors on behalf of the state. In order to perform this function (as well as others) efficiently, the Soviets must unambiguously own property that would yield them a regular income and must command a definite portion of the budget revenue. So far this is not the case. The paradox is that leasing is either not developing or doing extremely poorly because of the absence of market relations, while market relations stagnate because of unsteady development of leasing.

Third, the enterprise cost-accounting is approached as a leasing system (shop-sector-team) with defined ultimate performance results of each link and payment according to these results. Under public ownership of the means of production, a self-managed and self-financed enterprise is a "nourishing ground" for the development of a contractor and leaseholder economy. In any case, beginning from 1963, the numerous attempts to further team and shop contracts and leases were a complete failure.

And the other way round. If the workers are not granted a real co-owner's status in their "primary" collective where they do the work, it is then easy to roll down to a view of dependence and an attitude of rentier in relation to the enterprise and society. It is different when the earnings of each doer depend on the labor contribution to the ultimate product of his primary collective and thus of the enterprise in general. In this case, return to dependence is possible.

Fourth, the property insurance of enterprises and organizations against natural hazard, premature wear, loss of transport damage and other risks with a broad spectrum of insurance terms and conditions, up to full indemnity of the losses incurred. At present, the State Insurance Agency insures the property of collective and state farms shifting the compensation for the losses of public enterprises and organizations on to the state budget and banks. Till the beginning of the 1980s, the insurance of cooperative enterprises alone was thought right and proper. Apart from the state farms, the new process should be extended to public cost-accounting enterprises, associations and organizations of other branches. Otherwise self-financing will be wanting. The insurance terms and conditions for agriculture must also be reviewed.

All this could largely be achieved by withdrawing the State Insurance Agency from administrative subordination to the Ministry of Finance and establishing it as an independent cost-accounting organization conceived to prevent and minimize losses in the economy due to natural hazard and other contingencies. The functioning of cooperative, joint-stock and other insurance companies would be also useful. In the capitalist countries, no other businesses are as rich as insurance companies, and they invest heavily in the financing of the economy.

Fifth, the adopted procedure of placing state orders with the enterprises for the production and delivery of commodities can be regarded only as a temporary, transitional measure. They retain the old content in a new form and actually are the same administrative tasks renamed, i.e. without mutual liability. State orders were extensively applied in the Soviet industry of the 1920s. The People's Commissariats for Railways, Military Affairs and other public departments developed their relations with the producers on a contractual basis. The supplier undertook to produce and deliver, and the other party to accept and pay for the specified product at a set price. The economic relations of trusts and syndicates were made following similar principles of sale and purchase. State orders were however accepted and fulfilled as a priority and in some cases at a higher price. But that was their only difference from other orders. It was natural: a self-financed enterprise and administrative distribution of state orders (particularly with their present high share of the output, constituting between 30 and 100 percent) are like genius and villainy – two incompatible things.

The past experience can be of good service to the new management mechanism over the period of temporary "co-existence" of old and new methods. If the State Planning Committee and the State Committee for Material and Technical Supplies with their product distribution plans, a traditional pattern for them since "War Communism" (instead of sale and purchase plans), act always as cautiously in transition from the supply of enterprises by quotas and warrants (by "rationing", as academician V.S. Nemchinov wrote 25 years ago) to wholesale in the means of production, the self-financing and self-management of enterprises will be frustrated. A seemingly organizational and economic issue thus grows into a socio-economic one deciding the fate of the reforms in general.

Finally, sixth, the inertia of wage levelling introduced in the previous decades is a most heavy drag on the growth of efficiency and quality in the present situation. Such levelling can be overcome only when the wages fund of enterprises is formed from the real income after selling the commodity produced. To date this has been guaranteed in advance, which is, in fact, one of the manifestations of the "spending" management mechanism: the higher the consumption of resources, the higher the returns. It is immaterial for the remuneration of labor whether the produce satisfied social needs, accumulated at warehouses or was good-for-nothing. Not so long ago, the auditors from the

State Bank of the USSR found a large batch of shoes at a footwear depot in the North Caucasus that were made in 1954. It will hardly find a buyer, but the wages (even bonuses!) were paid to shoe-makers, tanners, machine-builders, transport workers and power engineers (and to those who stored them for more than 30 years). It is not just a "spending" mechanism, but direct evidence of embezzlement in management.

It was established during inspections at the enterprises of different industries in Moscow that labor productivity grew slowly and even ceased to grow (because of increasing absenteeism, undiscipline, heavy drinking) with workers at an age of 35 to 40 – in the prime of life and professional skill. The matter is that according to the planned ceiling the highest wage category is usually reached by that time: they just cannot earn more. The losers are not only the production workers, but society in general. It would be most expedient to extend the traditional (introduced at the end of 1950s by the newly set up State Committee for Labor and Social Issues evidently to simplify its own work) six-category rate scale (an eight-category one was in effect till then) to twelve or even more categories and to lift the wage limitation. A person must always have a prospect for career development and correspondingly higher earnings, plus the confidence that society is aware of, appreciates and encourages this growth. At the higher educational establishments we have a "scale" of only four categories: assistant, senior teacher, associate professor and professor. And what about vocational and comprehensive schools?

The enhanced role of wages as a stimulant must be linked with the incentives of economizing in both human and machine labor. At present, benefits of economies in labor remuneration remain in their entirety in the collectives, while the economies made in labor in the past go partly into the bonus fund. The overall size of bonuses for the fulfilment and overfulfilment of the plan is several times larger than that for the saving of raw materials and energy. And the material costs account for three quarters of the product fabrication cost. The degrees of stimulation thus happen to be different, and the result is that the attitude to the amount of materials supplied and consumed, and to their efficient utilization is no more than indifferent.

5. Workers as Co-owners of the Country

The question of the factory workers' status as co-owners of industry dates as far back as Robert Owen's experiments in the 1820–1840s. Incidentally, the socio-economic theory of Marxism and its conclusion about revolutionary transition to collective ownership of the means of production in society at large was conceived on the basis of these particular experiments. In it, the workers were regarded as co-owners of the socialized means of production.

Lately, current affairs analysts have often written about the diversity of socialist property forms such as public, cooperative and individual ownership and about lease, contract and shares. Diversification really can take place. The main aspect of property relations remains however outside the field of vision: how to change and where to develop the relations of nationwide ownership. Should the property be distributed to collectives, or in other words, eliminated? Should it be dismissed as of no significance? Should it be divided among the republics? Should it be left in its present state?

The whole point is that the work collectives will not become full co-owners of their enterprises until all workers across the country become actual co-owners of the socialized means of production. In view of this, two directions should be pursued in the present situation: democratization of the state economic bodies and search for realistic ways to involve in management the work collectives and the workers as co-owners of the means of production. Of course, some experience, quite limited though, has been gained in both directions. Proceeding from it and adapting it to the new conditions we must devise measures dictated by the very principle of society's socialist system.

Under the self-management and self-financing of enterprises, the stand of management and planning bodies acquires exceptional significance. As administrative organizations, they were for decades accustomed to "feel, think and act" on behalf of others. They will certainly try "to hold and to forbid", while they can apply mainly administrative methods. Placing them on a different footing of mutual relations would mean re-orientating them toward new levers of management and new logic of economic ties.

An administrative organization employing economic methods would be the same as having an air force regiment commanding the Aeroflot airline or the railways. Demanding the lifting of the "petty tutelage" of enterprises without changing the rights and liabilities would mean demanding the impossible. At present it is virtually unrealistic to control absolutely everything to the last trifle "from the center" (the ideal of the 1930s). Such ambition alone can choke the idea of cost accounting – the economic relations of independent and equal partners. The economic methods can be realistically and responsibly employed by an enterprise or an organization with its "own property", cooperative or leasehold property.

Incidentally, when the present branch ministries institute the stable norms of profit distribution to enterprises and associations, they run beyond any limits of liability. Generally speaking, the initiative does not originate with them. The procedure is established and imposed on the industries by the Ministry of Finance. Neither does the Ministry of Finance operate independently. The State Planning Committee adheres to the principle of planning "from the achieved level on". Money is correspondingly found in the economy on the same principle by financial agencies. Usually, everything boils down to the notorious "fact":

taking more from better performers and less or nothing at all from poor ones. It is a graphic display of the directionary procedure aimed against cost-accounting.

Of course, it will be more appropriate, consistent and easier to establish by law the tax relations between a cost-accounting enterprise and the state budget – the size of income deductions as determined by society. The taxing should proceed from a progressive rate scale because of the broad profitability gaps between the industries and the enterprises. Stability is then guaranteed by law and can be changed in the same way if required. Besides, it would be most appropriate to borrow taxation practices from the advanced countries where the income share earmarked for capital investment (new factories, renovation, retooling) is exempt from taxes. This stipulation encourages enterprises to invest in the improvement or expansion of production and not to spend uselessly. They work for the satisfaction of social needs without "pressure" from above. The individual norms of profit distribution are not norms in the proper sense: they are the traditional unitary tasks "squeezed" into standard rules.

In general, if the sectoral (inter-industrial) bodies of management acted as self-financed associations, corporations and firms, as "holders" of the main pro-ductive assets, and leased them to the collectives of enterprises "existing" thus on their own earned income, they would practically fit the market mechanism of management. They would interfere in the affairs of cost-accounting enterprises in cases specified by law (e.g., chronic unprofitability). The main area of their activity will shift then to research, design of original machinery and technology, radical renovation of enterprises and building of new ones for leasing. Each party would be responsible for its own field of work.

The State Planning Committee was set up by the decision of the All-Russian Congress of Soviets in 1920 simultaneously with the Council for Labor and Defence and, along with the Supreme Council for the National Economy, People's Commissariats for Railways, Communications, Food and others, was subordinated to it as one of the central economic departments. Its purpose was to work out scientific development plans for the national economy and instruments their implementation. Responsibility for conducting affairs was conferred on the executive departments which were also in charge of the resources. General guidance was effected by the government and the Council for Labor and Defence as one of its organs. Concentration of both the planning and management within the State Planning Committee in the last prewar years was due to emergency circumstances. Over almost half a century this became not only customary, but also accepted as the only solution possible, which is wrong, of course. Neither is the monopoly position of this organ (its status superior to that of other economic departments) in the spirit of the time. It is enough for the Planning Committee to retain its old style and the methods of work and the entire reform of management can be actually derailed. So far it shows no signs of impending radical changes. Only democratic control on the part of the working people as co-owners of the means of production can secure the badly needed changes.

I am unaware of any works concerning the real forms of co-ownership by workers in socialist society. The first such form is evidently defined in the economic reform's clause dealing with the procedure of nationwide discussion of draft plans. This economic and legal practice must certainly be included in the system of measures to be worked out and implemented. Such a sole measure is detrimental at least in that it becomes effective at the end of the planning process and does not involve the workers and their collectives in the process itself.

As to democratization of the procedure of decision-making the following seems most appropriate.

Equal rights (and equal liability) of the country's central economic organs – the State Planning Committee, the State Committee for Material and Technical Supplies, the Ministry of Finance and others – should be recognized as a general prerequisite for democratization in the management of the socialized economy. So far the Planning Committee forces them to submit to its preferences entirely "in kind" – the build-up of output in units, tonnes and meters. Complying with its decisions they even allot money for spending. The planning body thus has no need to "count money", to look for and to find the most effective economic solutions. It follows that neither are the latest technologies, techniques and management methods imperative to it. Instead of high-technology plants it can plan several ones with technically obsolete, conventional and backward equipment. It does not care about costs at all: he who needs will find money, it is his concern.

Effective spending can be done only by those who plan, and vice versa. The commanding position of the central planning body with its priorities "in kind" force the fiscal departments to activate all possible sources of financing the planned expenditure, including spontaneously developing inflationary ones. Under this unequal situation of the above departments, it is impossible to stop these spontaneous processes and therefore to balance the economy and to arouse interest in scientific and technological progress. Hence the inevitability of administrative price-setting: there can be no alternative in a dramatically and long disbalanced economy as there can be no scope for market relations.

Will the departments be able to agree on a single version of the plan as equal partners? And why not? By the way, a centuries-old practice can be borrowed from the Roman Catholic Church. As is known, the Pope is elected by cardinals who are locked into a room and kept there until they make a final choice ... Only a version that takes into account everything, probably even contradictory requirements, can be optimum or approaching the optimum: the "kind", cash in circulation, the dynamics and relationship of prices. What matters is that the plan must have a definite target and pursue social objectives whose achievement it must ensure. This is what the planning departments have in common and what will bring them together.

In order to coordinate the activity of all central economic bodies the present reform envisages the formation of an Economic Council consisting of top executives from those organs and headed by the Chairman of the State Planning Committee. Could a better decision be reached? Will it be conducive to the equality and equal responsibility of the departments? How will the decisions be adopted – by vote or by the chairman? If it is a council, the adoption or non-adoption should obviously not be decided by the chairman. If it is, we cannot talk about a council but only its substitute in form. The first steps of the new institution show that another advisory body where opinions are exchanged has been set in motion. And that is all there is to it.

Direct and immediate links between the provincial economic councils and the local bodies of Soviet power were one of the democratic aspects of the industrial management system established in the first years after the revolution. The reforms of 1930–1932 responsible for the economic mechanism of forced industrialization ended the influence of the organs of Soviet power within industry.

Since the first months of Soviet power the congresses of the republic's economic councils were regularly convened to outline, on a democratic basis, the current and, particularly, long-range tasks and methods of dealing with vital problems of industrial management. Two such congresses were held in the crucial year of 1918. The trade unions which were going to take the management upon themselves were also active in these directions. The second platform of the Communist Party (1919) envisaged the transfer of the national economic management to trade unions: "the organizational apparatus of the socialized industry must rest first of all upon professional unions. As participants, according to the laws of the Soviet republic and the established practice, in all local and central organs of industrial management the trade unions achieve actual concentration in their hands of the entire management of the entire national economy as a single economic unit" [26]. It is noteworthy that under Soviet statehood the transfer of economic management to a public mass organization was raised as a programme objective. Such is the path of transferring the co-ownership function to the working people.

At present, the co-ownership position of the working people could be practiced in the form of national, republican and local congresses of the councils of work collectives. The decisions on socio-economic development at an enterprise are subject to the general meeting of workers or the conference of their representatives and to the elected council of the work collective, and the management is accountable to it. Nationwide, this process could find

---

[26] The CPSU in Resolutions, Moscow, 1983, Vol. 2, p. 83 (in Russian).

embodiment and completion in a national economic council of work collectives authorized to discuss constructively and to consider democratically major social and economic problems (preplanning outlines, draft plans for economic development; social, investment, financial, structural, supply and pricing policy for the planned and long-range period; elections of economic managers, drafts of social and economic legislation, etc.). Under such nationwide consideration and control, multi-version planning for the purpose of choosing the best one could become a reality, the same as the examination of alternatives in choosing the methods of tackling the main socio-economic tasks in the course of fulfilling the plans. This is also democracy in management.

It is most important for the new form of workers' participation in management to become a "working association" and not a news or "discussion" agency. Substantiated decisions or resolutions of the congresses of the councils of work collectives could be submitted for discussion and adoption by the appropriate Soviets of People's Deputies, including the Supreme Soviet of the USSR, as organs of Soviet power. In the first case, the citizen will be represented as a worker and in the other as a consumer. The final decision rests with the latter.

The congress of people's deputies, beyond any doubt, represents our society in all its complicated structure – social groups and classes, nations and nationalities, ages, occupations, interests and much else. Each voter shares something in common and has something of his or her own – he or she is a citizen of the USSR and a resident of a region, town, district, a communist or non-party person, a trade-union or Komsomol member, a pensioner or inventor, journalist, a women's council member or a military man. And in each of his or her "personifications" the voter will be represented in the supreme body of power by deputies.

First and foremost, though, the voter is a worker, and a member of a work collective. But the representation of each of us as a worker is rather poor and even selective. The architects, journalists, designers, innovators and inventors, and others united in unions delegate their deputies to the congress. The professional sportsmen are represented by three deputies, the philatelists by one. And what about the work collective?

The work collectives have a right to nominate candidates and to take part in the preparations for and conduct of elections. Article 7 of the Law on Elections says to this effect: "the preparation and conduct of the elections of people's deputies of the USSR shall be carried out by election commissions, work collectives and public organizations openly and in public". Further on, the rights and duties of election commissions are clearly defined and the representation of public organizations at the congress of people's deputies is intelligently regulated, but not a word is said in the law about the role and place of work collectives. A democratically formed supreme body of power will hardly be properly representative without direct representation of the leading constituency – the work collectives. It is deputies from the work collectives that should make up a significant part of the deputies' corps from public organizations.

The new Law on State Enterprise and Cooperation in the USSR grants the work collectives numerous rights, including self-management and self-financing. Although the process is slow and difficult, its direction is clear: the workers of each collective elect their organs of self-management – the councils of work collectives.

Socialist society is a society of self-governed work collectives and a society of people's self-government on a country-wide scale. If so, a question arises: why shouldn't the law on elections envisage higher representation of work collectives at the congress of people's deputies of the USSR?

It can be argued that the workers, collective farmers and employees delegate deputies from their territorial and national-territorial districts. That is true. They delegate their deputies not as workers, nor as members of work collectives, but as citizens, as residents of the district.

In the press and at the 19th All-Union Party Conference proposals were put forward to convene the congress of representatives from the councils of work collectives. Its delegates would represent us all as workers regardless of profession – builders, architects, cattle breeders, and so on. Such a congress could discuss and settle all those questions considered within collectives on a nationwide scale – draft plans, restructuring of management – and could also propose socio-economic legislation. At such a congress, the people's deputies could be elected from the leading sector of our society – the work collectives.

This procedure will bring together the aggregate worker represented by the councils of work collectives and the Soviets of people's deputies. The state economic bodies will then become organs of the councils of work collectives – advisers, consultants and executors of tasks, and will be accountable to them. In practice, the idea of socialist society as a nationwide cooperative is being implemented. The management mechanism will then correspond to the nature of the social system. The country's workers will indeed become the aggregate owner of the public means of production. Democracy will also embrace the basic sphere of society – the economy. The humanistic orientation of socialism will acquire its natural basis.

6. Priorities

Apart from strategic direction, the reform also needs a working programme. In the present situation, at least the following priority measures are imperative.

The first is to curtail the number of projects under construction. We have over 311,000 of them in the country. On average, 12–15 workers are employed on each site. Small wonder that they take unbelievably long to build: 8–12 years instead of two or three. In 1988, the builders fulfilled the plan of contract work to the extent of 100.2%, but the basic plan of commissioning production capacities only 77%. About four billion roubles were paid to them in wages for the growth of supplies on construction sites. Above 150 billion roubles are frozen in such projects for a long time. Taking 10 years to build a factory at the end of the 20th century is equivalent to losing everything. It has nothing to do with scientific and technological progress. For instance, the modernization of over ten shoe factories on the basis of advanced technologies was commenced in the past few years. The question is: when will they start producing? At the end of the century? Besides, ever larger portions of the state expenditure are redirected into social needs – hospitals, schools, clubs and such like. This time the question is: can we wait 10–15 years for them to become operational? Of course, we could give the "green light" to these particular projects and speed up their completion. But what about all the others that were started long before and have devoured billions upon billions of roubles? They are a heavy burden not only on the construction industry but on the entire national economy. The stock of uninstalled equipment alone increased from 12.6 billion roubles in 1987 to 14 billion roubles in 1988. The number of new projects increased by 53% in 1988 and their overall estimated cost by 56%. The means frozen in unfinished projects are the main cause of the critical financial situation in the national economy. Naturally, if the money fails to return to the treasury in due time after marketing additional output, printing presses have to be set going, which constantly overflow the circulation with billions of excess roubles.

In order radically to raise the returns from capital investment, to speed up scientific and technological progress and to promote a healthier rouble we shall evidently have to risk a surgical intervention – closing down for the time being or even forever of a considerable number of the unfinished projects, maybe half of them, maybe more (this is yet to be decided) to complete and put into operation the others as fast as possible. It is an extraordinary measure, but not a new one. At the turn of the first and second five-year plan period when the number of enterprises under construction exceeded all reasonable limits, this kind of "financial manoeuvre" (as they delicately put it at the time) was carried out, which certainly yielded positive results. Even before, at the beginning of the New Economic Policy, production was concentrated at technically better tooled enterprises to intensify the operation of the almost idling industry. In Petrograd, for example, in the first half of 1922 there were 291 factories operating, 435 had been closed down and 90 were leased to private enterprises. The industrial output growth amounted to 38.3% a year.

The second measure concerns the State Bank – it must pursue a tough credit policy aimed against inflation. All the money circulated (cash and cheque book money) has one source – the vaults of the State Bank. When the bank grants a loan, the amount of money in circulation increases, and correspondingly decreases when it is paid off. The whole trouble is that the bank is not yet the architect of the national monetary policy. It has to obey the decisions of the State Planning Committee which is not liable for this area of management. It is like subordinating road traffic patrol to the motor industry. The results are sad: the banks are forced to write off the debts of enterprises and to grant loans to those who will definitely fail to meet them because of poor performance. Unless the money is duly returned to the bank, the printing press must be repeatedly put to work. On the whole, the situation is such that for many years the lending to enterprises has twice outstripped the growth of output. Regardless of other reasons, the abnormal situation is thus predestined: the buying enterprises and the population always have more money than there are commodities at the producers' storehouses and supply depots.

About 100 to 120 billion roubles of current assets have accumulated at the enterprises in the form of excess and needless reserves of raw materials, semi- and finished products. Those are dead stocks inhibiting effectiveness. There can certainly be no serious talk about self-financed enterprises, because nearly all of them exist to a higher or lower degree at the expense of banks. The economic levers of management are again ineffective, because they are actually financial ones: the norms of profit distribution, wage and bonus funds, depreciation of fixed assets, fines, etc. Neither can there be efficient pricing: if the available money deliberately surpasses the availability of commodities, the prices always go upward and they must willy-nilly be kept down by administrative control, which is not at all failure-proof. How many inspectors must be employed to follow up the prices on about 24 million items?

The conclusion is that we cannot do without a tough credit policy. It means that the credit growth rate must continue to be lower than the output growth rate with every passing year. A cost-accounting enterprise must realize that it can prosper only on its own earning, economies, fast turnover and liquid assets. A bank must not act as an omnipresent magic wand, but as a source of help at times of real need. This should be the crux of the banking reform.

New banks are being set up by the industries, cooperatives and joint-stock companies. A vast country would certainly need many banks. Nearly all the newcomers however wish as great independence of the State Bank as possible. Independence and responsibility are necessary and obligatory. But the monetary system of any country is integrated. Nowhere in the world is there a bank independent of the country's central bank, of the State Bank which alone sets the rules and issues money. Any commercial bank can act at its own discretion within these particular limits. Definite rules of the relationship between the State Bank and all the others are therefore needed.

All banks must keep their accounts with the State Bank, otherwise the country's monetary economy can run out of control. Only then will the State

Bank be able to pursue a uniform monetary policy. A commodity–money economy cannot survive without it. Another trouble is that the status of the State Bank in the country is lower than that of the State Planning Committee. If this evident fault is not corrected, the role of the banking system will not be enhanced in the economic management and, consequently, inflationary tendencies will continue in our economy. It is impossible to balance the economy and to accelerate its proper growth without settling this key issue.

Lastly, the third measure which concerns national economic planning. It is a priority question for a planned economy. Suffice it to say, that to preserve the traditional procedure of state planning, unchanged for nearly six decades, is practically to torpedo the economic reforms. Indeed, the plan comes first of all. But a few months after the adoption of decisions on a radical reform in summer of 1987 the words "the reform of national economic planning" somehow dropped out of our vocabulary. At first they were drowned by "self-financing", "price reform" and "wholesale trade", a bit later by "co-ops", "lease" and "share-holders" and then it was as if there had been no decisions on the reform of planning. It was quietly withdrawn from verbal usage.

Neither was it included in business usage. The State Planning Committee is approaching the next five-year plan with the same old methods tried out and polished in the past. Shortly speaking, it is a system of plans for the production and distribution of produce – steel, oil, fabrics, grain, footwear, etc. The only difference is that the previous system comprised above a thousand such plans, while the new one will have about a hundred and fifty. No practical changes. It does not even smell of radicalism there. When will it reach the Planning Committee?

It must reach the committee by all means. The matter is that the present method of economic planning dates back to the War Communism period and the famous plan for the electrification of Russia – GOELRO. Money, finance, effective demand and commodity supply then played no part in planning and management. The count was in units, poods (36 pounds or 16 kilograms) and versts (two thirds of a mile or 1100 m). Enterprises deprived of any kind of independence fulfilled the orders from the "center" without fail: they produced and delivered. Some changes were introduced in the first five-year plan periods, but the original orientation remained: more units, tonnes and meters. At present, the situation is changing radically. The commodity producers are no longer obedient enterprises but zealous merchants. It was so declared by the law on state enterprise, the law on cooperation and other legal acts of the present reform. In this situation they will not have the "center" speak "units" language to them, let alone command.

Either the plan will "talk" business in plain words with the cost-accounting enterprises or cost-accounting will be suppressed and the enterprises placed under the administrative command system anew. It is either one or the other.

In any case, it is no longer possible to build a new management mechanism on two mutually exclusive foundations – cost-accounting on the level of enterprises and "in kind" on the level of the State Planning Committee. There can be no doubt about it. We remember the experience of 1965 and 1979 when we tried to introduce cost-accounting, but failed to change the planning so that both times the result was the same – the cost-accounting shrank and withered. Even not so long ago, at the beginning of the 1980s the same was repeated under the so-called large-scale experiment in the industry, no longer remembered now ... Isn't that enough?

It is time that the national economic planning be raised from its initial and simplest forms and methods to a qualitatively different status. In fact, the first real planning reform after the GOELRO and first five-year plan must be carried out. It is a decisive prerequisite for success of the economic reform as a whole. A resolute transition from the system of demand and supply plans should be at the heart of the matter. The task is not so difficult. Instead of the "produce and deliver" command the plan must predict the demand of enterprises and the population for different commodities and envisage how to meet and to control it by means of capital investments, structures policy, pricing, credit, tax and all other instruments of economic management.

7. Conclusion

The theoretically derived general law states that socialist society is a society of self-managed economic collectives – from primary units to society as a whole. "Actual possession of the whole industry, all the implements of labor on the part of the working people" (Engels) is its natural material basis. Socialist society must be "one large cooperative", "a system of civilized cooperators", "one national or, more exactly, nationwide cooperative" (Lenin).

In the USSR as in the other socialist countries, it so happened that formal nationalization was not followed by actual possession. The development of the socialized economy (under formal socialization of the means of production) did at first speed up economic and scientific and technological growth, mainly due to extensive factors. This economy as a whole nonetheless displayed its inefficiency and is displaying it ever more, first and foremost because of the formal approach to socialization.

Real socialization implies that a worker and all the workers in the aggregate act as joint owners of the means of production and are responsible for and interested in making and carrying out only reasonable and effective economic decisions. But this was not the case. As a matter of fact, this society has been formed without owners of the means of production, and acting as owners are the managers and supervisors from team and shop level to the State Planning

Committee and the Council of Ministers. A manager can be good or bad, but without constant "pressure" from the owner he will naturally care first and foremost for his personal interests – to have his superior satisfied with him.

As a result, the alienation of workers from the means of production which theoretical socialist thought had ascribed exclusively to exploitative socio-economic formations spread, in practice, to Soviet society, too, and manifested itself in the most complete and finished form. In those societies, individual and collective (share-holding) owners were active who had to pull efficiency upwards by virtue of their actual position in the economy. Incidentally, for this purpose they practiced material or moral encouragement of both the managers and through them the workers. Under socialized ownership the real owners of the means of production have not yet materialized. In view of the above observations, Soviet society can be characterized as elementary, primitive socialism.

For this reason, a system of consistent measures for making all workers the real aggregate owner of the means of production should be regarded as the strategic direction for the development of the management mechanism. The economic collectives must be made co-owners of enterprises who will conduct business on the basis of leasing the public means of production, on the basis of their own means of production (co-ops) or as state-cooperative enterprises. As to individual enterprise, it will evidently develop (with a certain number of exceptions) into different cooperative enterprises (marketing, procurement, crediting, etc.).

Neither was another thesis of the theory of socialism confirmed in practice – that of the non-commodity, money-free character of the socialist economy. The experience of all the socialist countries shows the immanence of commodity–money and market links in the economy throughout the 20th century. All these years state power could restrain, "oust" and drive out the market relations with varying success, but could not put an end to them in the economy. Even in the years of the fiercest struggle against the market, under War Communism, they did exist and embraced tens of millions of people in Russia. Apart from money and market relations between the producers and the consumers, civilization has created no other general index of costs and management results. Only its forms changed ranging from gold holding to "electron" money. True, at a certain stage, even the development of a form can lead to the change of content. Today however this question can hardly be regarded as pressing.

With the appearance of national economic planning and up to the present, the "in kind and by piece" structure of the plan for social and economic development has been assumed as an axiom, which blocks the road to the market economy. Since the plan is a key element in managing the socialized economy, an insoluble situation arises with economic planning "in kind". It is impossible to have a plan based thus while cost-accounting and self-financing apply to the enterprises. In such circumstances, the "in kind" always takes the upper hand, while cost-accounting is emasculated and destroyed. The economic reforms beginning with the New Economic Policy and ending with the present radical one gravitate toward the same result.

The traditional basic instruments of national economic planning, the production and distribution plans, have become hopelessly obsolete. The attempt at transition to a more advanced procedure of planning during the first five-year plan period (coordinated programmes for construction, production and finance) was frustrated, which made us roll back to where we were and stay there for a long time. Planned socialist market economy is feasible only where plans for effective demand and commodity supply are the principal instruments. They alone can ensure the unity of the whole management system – the compatibility of money, volume and price levers. At the same time they "speak" a common language with the enterprises and organizations operating in market conditions. They also make real the rejection of the mainly administrative methods of management. Otherwise another return to the initial management "in kind" on the basis of socialized ownership is unavoidable.

In its developed form, the socialist economy will be a planned market economy of associated producers. It means: first, delimited circulation is underway of productive assets, labor products (commodities) and returns of the interacting self-managed economic collectives; second, the collectives join self-managed associations (marketing, R&D, fiscal, regional and others); third, they all make up a united economic association organizing, planning and managing the production on democratic principles on society-wide scale.

This association in the person of effective and regularly renewed representative organs is accountable for its activity only to the country's supreme bodies of power – the congress of the people's deputies and the Supreme Soviet. The latter represent the workers as citizens and consumers. The structure of rights and obligations in society subordinates the aggregate producer to the aggregate consumer. Depending on the subject of their activity, the specialized economic departments are subordinated either to legislative power or to the economic association (it could be, for example, the congress of the councils of work collectives), act only as qualified advisors and executors of tasks assigned to them and elaborate alternative decisions for consideration. In the present situation, the congresses of the councils of work collectives could reproduce the functions of the congresses of the national economic councils of the Russian Federation in the first years of Soviet power which democratically examined the basic question of management (how to plan, whether to spend money, etc.). For all practical purposes, this will make true the principle "the socialist society as a nationwide cooperative". The real management will thus come into accord with the true nature of the social system.

The cost-accounting and self-financing of enterprises, associations and economic organizations make a natural basis for the self-management of collectives. The slow and largely formal development of self-management after the 1987 decisions is due not only to the lack of experience in collective decision-making, but also to many "lapses" in applying self-financing – in the main, to discretionary withdrawals and "injections" of financial resources by the superior (administrative) organizations. A tough credit policy is needed (as well as a tough issuing policy of the State Bank at the center) and withdrawal of profits to centralized funds exclusively through taxing, with contractual forms of relations between the suppliers and the buyers.

As to cost-accounting relations on a regional scale, the self-repayment of the economy of the local Soviets and that of republican subordination seems imperative. The local and republican budgets must have their own fixed incomes as well as expenditure ensuring their autonomy and self-financing as full-fledged participants in economic activity. On occasion they are entitled to repayable or grant aid from the state budget. The republican and local Soviets can act as users and managers of the property needed to perform their constitutional duties concerning housing, public health service, land utilization, environmental protection, etc. So economically it makes no difference to the central and branch organs where to site new production capacities, either in the Siberian tundra or the Fergana Valley, where the labor force abounds or is lacking. The local authorities can do little to change anything. The compensatory requirement will make the industry and the regional partner understand that what is advantageous for each of the parties is also advantageous for society as a whole. Each partner has his own budget, his "purse". If a local Soviet provides education, health service and housing from its own budget, the department or enterprise recruiting the work force must pay the Soviet to compensate for its expenses. The scientific principles of evaluating and establishing rates for the use of all kinds of resources have been long developed and approved. The use can already be charged. The regions will have to ensure self-repayment of the economy within their territory and will thus become interested in and responsible for efficient utilization of the resources. The economic incentives will then encourage them to increase the public wealth.

The principal task in intensifying the socialist economy is clear, but not at all easy: making the worker a real co-master at his enterprise and co-owner of the public means of production. All practical steps and measures in restructuring the economy must be viewed from the pro-and-contra standpoint. The respective was precisely defined by Lenin: "No one will be able to bring ruin on us, except our own errors" [27].

---

[27] V.I. Lenin, Collected Works, Vol. 42, p. 249.

Chapter XII

Efficient employment and the labor market in the USSR

Irina S. Maslova

Doctor of Economics, Institute of Economics, USSR Academy of Sciences

The development of market and commodity–money relations in the present-day Soviet economy brings about substantial changes in the employment mechanism, and gives rise to a great number of social problems while their solution requires a new philosophy and new theoretical approaches.

The formation of a labor market and the readjustment of the state system of manpower management for efficient employment requires the institution of legal and material protection of the citizens against the possible risks involved in the country's transition to a multistructural economy with free competition. It will not come by itself and needs an active yet flexible social policy.

The present-day Soviet economy is characterized by a new problem alongside the chronic unsolved ones; the aggravation of employment problems.

There is a marked contradiction between economic and social criteria for efficient employment.

1. New Developments and New Problems

Analysis has shown that negative tendencies in the use of the labor potential still hold in the present-day employment situation. A high degree of involvement of the able-bodied population in public production goes hand in hand with low efficiency in cost terms of social labor. All this does not in any noticeable way benefit the welfare of the majority of the population in this country. Optimization of the structure of employment and a higher efficiency of social labor is made difficult by overmanning in public production, high latent labor reserves within an industry, as well as low mobility of personnel.

In the years of "perestroika" there have emerged a number of new tendencies in the employment process, with diverse impacts on efficiency. Labor mobility has become greater. On the one hand it is a result of a more active redundancy process at the existing enterprises due to their transition to self-sufficiency and self-financing, as well as of changes in the structural and investment policy, management cuts, closing down of wasteful production, and the disarmament policy. On the other hand, transition to the multistructural economy and competition among its sectors made some of the workers move from the state-owned enterprises, institutions and offices to alternative production spheres. The state-owned enterprises started experiencing an acute shortage of skilled personnel, since the skilled workers (in the age range 28–42) left for the cooperative sector of the national economy. Working has become more difficult for women with small children, for school-leavers, graduates of technical secondary schools and college graduates with no practical experience, and for the physically handicapped, former prisoners, workers dismissed for their conduct or absence without leave, those of preretiring age and demobilized soldiers.

Greater disproportions have arisen between labor and job supply and demand in the new development regions where there are added difficulties with new directions, intensity and structure of migration flows, and smaller volumes of regulated territorial redistribution of labor.

There have appeared vast areas with acute local employment problems, namely: old "no-hope" industrial regions, regions with mining and timber industries and mass-scale use of seasonal labor, naturally inhospitable and ecologically degraded zones; densely populated parts of Central Asia, the North Caucasus, Azerbaijan, and some others. Sometime this provokes social tensions, inter-ethnic conflicts and job competition.

Lifting emigration limitations both for permanent residence and for contract work with foreign companies, increased immigration of labor from the countries of Eastern Europe and Asia accounts for some new developments in the situation. Migration leads to the drain of skilled labor and has a negative bearing on the quality of the labor potential, because specialists are generally leaving, to be replaced by unskilled labor.

The list of new problems could be continued. But it is essential to point out that the most radical changes in employment are due to the revolutionary readjustment of property relationships.

The Law on Property in the USSR passed by the Third Congress of Soviets in 1990 changes the conventional concept of socially profitable labor, and the terms, forms and principles of employment. Labor other than using state-owned or cooperative means of production is now also considered as socially profitable. It is now a law that people who work in kitchengardens, on a farm or elsewhere are doing a socially profitable job, and enjoy the right to dispose of their ability to work (Art. 6.2) and the proprietary rights to the means of production they need to do the job (Art. 7.1). The Law on Property not only recognizes a variety of forms of enterprise but also defines the state's responsibility to secure the necessary conditions for their development and protection. Expansion of legal sources of income, proclaimed by the Law on

Property, is a key to the understanding of the strategic meaning of the changes of the principles of employment. At present it is not only the income that a person gets by participating in public production or his own business (Art. 14 and 17 of Constitution of the USSR). Due to the reform all investments in credit agencies, shares and other securities, as well as inherited property are recognized as a source of income. Universal and compulsory labor in public production is being replaced by the principles of freely chosen employment which reflects the right of every able-bodied individual to choose a production sphere and a type of activity he would like to participate in. Though hired labor is now permitted as a possible alternative (Art. 1.4) the Law on Property is essentially designed to protect the interests of working people. First, whatever the type of property, the exploitation of one man by another should be excluded. Secondly, whatever the type of property, a person is entitled to wages and the agreed working conditions, as well as social security, envisaged by the current legislation.

At present the idea of full employment is being attacked due to the wrong and vulgar interpretation of the term as participation of the entire able-bodied population in social production. Such understanding of the fundamental principle of socialism is invalid. The principle actually means the responsibility of the governmental bodies as representatives of society to provide everyone who wants to and needs work with a possibility to get a job according to his or her talents, aspirations, the type of received education and vocational training, with the social needs taken into account, and a minimum guaranteed wage. The principle corresponds to the right to work and receive education ensured by the USSR Constitution for its citizens. Proper employment is one of the means to implement these principles.

The principle of full employment does not deny people the right and opportunity to be engaged in socially profitable labor for an income outside the framework of public forms of enterprise, such as: housekeeping by agreement, tutorship, babysitting and looking after the sick and the old, and other paid services. According to the data of the State Committee of Labor of the USSR, in 1989 alone 4.3 million people, or 3.1% of all those engaged in the national economy gained income from outside the public sector. Apart from these people who do not claim jobs in public production, there are groups of able-bodied population who do not work during different periods of time for various reasons, some of them out of their own free will (approx. 9 million). The fact is that some of them would like to work and actually need a job to earn their living, but find it difficult to get.

Despite the mounting demands for officials to admit the existence of unemployment in the USSR, the current statistics fail to distinguish the unemployed from those temporarily out of work. A number of expert judgments have been made public, according to which the number of unemployed varies between 1.5 and 23 million. These figures are far from real and only mislead the general public and build up emotions. That makes it extremely important to define the status of the unemployed, to work out the rules of legal protection of the worker's interests when forcibly out of work, and to reimburse for the lost job and wages. It is necessary to improve statistical records of the unemployed in order to determine the real situation concerning job demand and the means to fill it, finances needed for social support and reimbursement for those who would be defined as unemployed.

In the practical situation of the USSR a person can be qualified as unemployed if he has lost a job or any other source of income not through his own fault; if he is registered at the employment agency as able and ready to work according to his specialty or take a retraining course, and if the agency failed to offer him any job during the legally fixed period of time.

Elaboration of the measures aimed to enhance social security against unemployment in the USSR is made easier because the humane principles of solving this problem have been worked out by the International Labor Organization to suit countries with different social structures and levels of development. This fact, however, does not make the problem of readjustment of the social institution of employment to the general principles of the national economic management reform less important.

2. Building the Labor Market and Increasing Labor Mobility

Greater employment efficiency involves the development of the labor market and competition among enterprises for personnel of the required specialty and skills, and among workers for the jobs that would provide better working conditions and higher pay.

There is still disagreement as to whether a labor market exists in the USSR or whether it is a problem to solve in the near future. Only a few years ago the question was impossible because Soviet Economics would give only one possible answer: a labor market does not and cannot exist under socialism. These are the main arguments for the given conclusion: In society where public property prevails, a worker is not alienated from the means of production, he is a co-owner; labor activity is not treated as a commodity in this case and it is not alienated from the employee; the process of associating labor with the means of production is of a social nature and does not need a market, and its industrial and territorial distribution and redistribution is planned on the basis of direct economic relations.

The economic reform demanded that the conventional notions of what seemed clear and undisputable be reconsidered. In view of the latest changes in the economic strategy there has been a number of strong arguments to recognize the reality of the labor market in the USSR and its necessity for the effective performance of the Soviet economy.

First, in theoretical terms it is generally recognized that the prevalence of the command economy has engendered wide-scale alienation of working people from the means of production and from the free enjoyment of its results. Social injustice in distribution has brought forth a great number of poor people who have to work overtime to get extra money to improve the living standards of their families (e.g. paid services, or another part-time job).

Second, it has been recognized that the failure to take into consideration the Law of Value had adverse effects on the process of determining the minimum cost of living, payments and pensions enough for a person to live a long life and be in good health, to raise a family and lead an active working life. It has been condemned as erroneous to set production and employment proportions by one-sided centralized decision without a due study of the consumer's demand, the market and the role of competition in filling it, as well as of the mechanism of sectoral manpower distribution and redistribution.

Third, due to the transition to the multistructural economy and recognition of the various forms of ownership as being equal and oriented to their competition, an individual has a substantially greater economic freedom to choose where to work, what to do, and in what field, also to choose the working hours and the way to earn a living.

Fourth, a multistructural economy gives a start to a new model of economic management. That means that the functions of the centralized control should be taken over by the republican and local authorities; it will be enriched by economic methods of management, a wide use of commodity–money relations, development of a full market system (comprising the means of production, transport, securities, housing, livestock, etc.).

A denationalization policy and a transition to a multistructural economy and equality of the various forms of ownership undermines the monopoly of the central agencies to establish wages, and their rigid regulation of production proportions and employment. Zones with higher incomes and standards of life, and better chances for earning money are being set up, with free exit of labor to competitive employment areas with only market laws to regulate it. There is also a possibility to overcome people's indifference to acquiring knowledge and improving their professional skills, thereby raising their productivity.

In this way a new flexible mechanism is being built to maintain the proper balance, proportion and correspondence of the elements of production (labor and the means of production). It is based on the integrity of the two ways in which the economic interrelation of labor and various forms of ownership of the means of production manifests itself, namely regulation and market.

The above considerations warrant the conclusion that a labor market already exists, albeit limited and deformed. It operates to some degree in the matter of redundancy, and in cases involving mobility and migration, when people change job and residence looking for better pay and conditions, and a better place to live.

The labor market in the USSR does not yet play an independent role. Currently to some degree it ties individual needs and the interests of the economic center in charge of government orders, the structure and location of production, recruitment to the public sector, the provision of material incentives and attempting to retard the development of non-public sectors. Nevertheless there are enough grounds for the conclusion that we are currently witnessing an active establishment of market relationships aimed at a fusion of labor and the means of production, manpower distribution among various employment spheres with appropriate projects for personnel training and retraining. The study of manpower fluctuations and migration processes makes it possible to claim the existence of local, national, state, and international labor markets under the specific transitory period of passing to the new model.

In general the employment situation in the USSR in quantitative terms cannot be considered as acute. In some regions it is becoming worse because of the mounting shortage of labor at state owned enterprises. According to data published by the State Committee of Labor in 1988, there were about two million vacancies in four sectors of the national economy (industry, civil engineering, trade, transport). In actual fact, the number is much bigger. The problem of insufficient applicants and surplus vacancies is not of a local and structural nature, because the mobility of the able-bodied population is not active enough in some regions, while the regulated system of employment was historically oriented to the labor demands of the enterprises rather than to the interests of the people and this was typical of extensive economic development with its growing shortage of labor.

The labor market in different regions of the country has its distinctive features, in urban and rural districts, it differs according to the specialty of skilled and unskilled labor, according to sex and age. Serious errors in the location of plants and factories account for the imperfect coordination of demand and supply in the labor market, and their variation on the territorial level. Over a long time it was neglected that there are pronounced regional differences in the quality of labor potential and the degree of professional and territorial mobility of the employable population. There was a tendency to neglect the national traditions and affinities of the people living in different parts of the country regarding different types of activity, their different cultural and historic backgrounds, and specific way of life in the formation of the sectoral structure of production and in the establishment of economic relationships. There was a marked difference among regions in social infrastructure, housing facilities, commodity market, environment. The failure to take account of regional manpower and job demand factors, and of the people's interests in work and employment explain the diverse situation in the regional and national labor markets in the USSR. At the same time it is an evidence that the all-union labor market is rather underdeveloped.

It is not only the variety of ways of life and traditions arising from the multinational composition of the USSR that slow down the mobility of its population. It is further hindered by other factors, such as the rigid rules for domicile registration, a poor housing market, difficulties involved in changing a state-owned flat or moving to a new one, lack of retraining facilities, and a social psychology and legislation geared to tie people to the same enterprise and neighborhood, i.e. to stability rather than mobility.

As building of the labor market is a step towards productive employment, it has become necessary to start a campaign aimed to change the people's psychology and cultivate a number of qualities, which are going to be vital for the people to be able to exercise their exclusive right to work and to benefit from various forms of ownership and economic activity. The point is to foster in people an ability to move and change jobs, this being the only way to make the most of their capacities for productive and creative labor, to stimulate labor mobility and to increase the efficiency of social labor.

Another task which is no less important is to rid people of their psychological adherence to the public sector and of their orientation to lifelong ties with the same work team and the same working place. It is of vital importance that every worker should want to improve his skills, to look for the best field to apply his talents in and fulfil his aspirations, and to be prepared to take a retraining course, if the necessity arises. To ensure the ability to work throughout one's lifetime in the context of a highly competitive labor market one should cultivate the above qualities, as well as feel responsibility for the entrusted work, to display initiative and keep discipline. These qualities are indispensable for a nationwide labor market, productive manpower distribution and redistribution and a coordinated system of demand and supply in different regions of the country.

3. Building of the Regulated Labor Market

A search is currently under way in the USSR for a model that would regulate the employment process and the labor market in the republics and the center. Some republics could be grouped on the basis of common tendencies in the birth and death rates, natural population growth, the nature of migration processes, employment conditions and job security of the population. The models regulating such markets are unlikely all to be the same.

The task is made easier for the USSR due to the existing international practice of market regulation and the experience of countries with different sociopolitical systems. In view of the great variety of the existing labor market institutions and the high social price of miscalculations in the employment policy and its regulation, the Soviet state is fully aware of the responsibility of choosing the right model and the right strategy for labor market development. The model should be chosen on the basis of its correspondence to the principles of political organization, social justice, and highly productive employment envisaged in the humane social strategy of the USSR. The labor market is a derivative of the general model of economic management and its regulators should be integrated into the model and should fit it.

The model based on centralized command planning of the development of the national economy, and the market economic model have completely opposite mechanisms to regulate labor supply processes (training, distribution, redistribution, retraining and use). There are all sorts of intermediate models, based on the widely used indirect economic methods of economic regulation and rejection of direct planning of economic units.

Two versions of the model designed to regulate the labor market, which correspond to the two conceptions of the reform, namely radical and moderately radical, are being worked out in the USSR today. The first implies a transition to a free market economy with a view to a full-scale manpower market with all of its intrinsic functions orientated to the possibility of hired labor, unemployment, and a system of measures that would protect people in case of unemployment. Unemployment in this context is seen as a kind of lever that should turn employment into an economic category and make economic activity more efficient.

The second version is meant to build stage by stage a multistructural economy, to stimulate competition, to coordinate planning and marketing in processes of labor supply. The regulated labor market corresponding to the second model has a limited number of functions designed to "adjust" the proportions of training, distribution and redistribution of manpower, the essential reserves that are shaped by the policy and actions of the economic center, and the system of government orders, to the actual needs of production and consumer demand.

Each model covers specific and shared tasks of regulating the labor market. In the first case the main task is to overcome all the barriers that hinder the development of the labor market, that is, to lift limitations in developing various forms of enterprise, to change the conventional system of distribution, to bring the level of minimum wages up to that necessary to provide normal and healthier living conditions, to abolish the system of double distribution (by the results of work and from the social consumption funds), to lift the administrative migration restrictions (on domicile registration, immigration and emigration in the USSR, residence in any republic), to build a flexible system of selling and exchanging the housing property and to ensure free and adjustable wages.

The task of regulating the labor market, according to the second model, is to optimize its boundaries, to determine its influence on the distribution relations, to see that it maintains social justice, to harmonize relations between employer and employees in the labor market, to take timely measures to protect working people against the risks arising from the operation of market relations, and to provide the necessary conditions for individual fulfilment and employment.

Both entail the revision of the Labor Code in matters pertaining to hire and dismissal, legal protection in cases of bankruptcy, phasing out of wasteful enterprises, and sacking, and the working out of measures that should promote initiative, assure wide-scale and secure employment, with social guarantees of the right to work.

Economic independence and sovereignty of the Republics makes it necessary to re-establish, on a democratic basis, manpower supply and management and the procedures regulating employment processes and the labor market of the economic center, republican and local authorities. Some general priorities that cannot be solved by the republic or territories, need to be identified to cement the unity of the federation, and to ensure the interests of the republics as part of the integral national economic complex.

At the same time it has been recognized that republican and local manpower and employment management should become more independent to stimulate productive employment of labor according to the nature of an enterprise, to help with employment and retraining of everybody who needs a paid job.

An active employment policy is a major task of the economic center. To make all managers understand clearly that full employment means economically efficient employment is the first step in solving the problem. This understanding will break the ideological barriers that make manpower mobility and a labor market impossible. The management should fully understand that it is very important to build a high quality system of education, to provide better vocational guidance, training and retraining facilities, and to work out and conduct a vigorous personnel policy throughout enterprises.

The policy of the economic center should be based on the recognition that professional employment is the priority task in public production, since it is a decisive factor that influences the people's ways of life and their personality development. In view of this one of its functions is to define a long-range employment strategy based on the planning of social and economic development of the regions, location of the plants and factories, investment, technical, tax, credit and finance policies. The state is to take measures to increase the competitiveness of manpower in the labor market, to boost mobility of the labor resources and to meet the people's demand for jobs in public production without lowering living standards. It also shoulders the responsibility throughout the country for preventing unemployment by promoting employment of groups of the population with different social and demographic characteristics. It establishes a national service and a special fund to secure employment, retraining and assistance to citizens who have lost their job or who have difficulties finding a job. Everybody is entitled to a common minimum of social guarantees to exercise the right to work, vocational training, retraining, unemployment benefits and social reimbursements. Besides, any sovereign republic or enterprise can expand the corresponding benefits and reimbursements above the fixed limits at its own discretion.

Such understanding of the functions and responsibilities of the center is likely to decrease social risks during the transition period from the existing economic relations to intensive development, market relations and the socially necessary level of population mobility, as it is envisaged by the moderately radical conception of reforming the system of public organization of employment and building market relations. This approach lays foundations for a democratic redistribution of managerial functions designed to control the population processes, employment and the labor market without prejudice to workers' interests.

In this way the principles underlying the labor market ensure common human rights for every republic and their right to take independent decisions on employment and development of the population. The main problem lies in the fact that the ways to coordinate national and regional interests have not been elaborated, the functions of the center and the republics regarding budgetary allocations are not fully determined. There still exist different approaches to the role and content of the plan and it is still not decided as to what is the future of the social consumption funds previously considered as major levers in controlling employment and the territorial distribution and redistribution of manpower.

The following instruments can be employed to regulate the labor market and employment processes in the USSR: planning, standard rules for economic activity, hiring and dismissal, working conditions and payment, manpower engineering, taxation, and control over enterprises to see whether they abide by the accepted legal norms and regulations and system of sanctions. Greater importance is attached to the forecasts concerning future changes in the economic structure and the locations of plants and factories, the sectoral and vocational structure of the labor force, as well as the programmes for the mobility of labor within the Union, republics and regions in view of long-range tasks of economic development, and to regional employment programmes.

The internationally approved methods of regulating the labor market and employment processes will be widely used, namely: updating of the pension and social security laws to decrease the demand, development of the community services (on the material basis of the local Soviets of People's Deputies) in rural and urban areas, establishment of associations for support of small business (financial help, consultations, etc.), stimulation of the kind of business that involves working at home, as well as part-time jobs, temporary employment, and contract migration.

The state labor and employment management system is to be readjusted to adapt it to social needs, to revise the structure and the principles of the existing relations between the State Committee of Labor of the USSR and its counterparts in the republics and the Soviets at every level to give it more rights to regulate the employment processes, and money to provide social security to temporarily unemployed people. Due to the fact that the interests and needs of people have become the main concern of employment agencies, the center had to do a big and meticulous job to retrain their staff to meet the new tasks and requirements under the changed employment conditions. The focus is also on the technical re-equipment of the regional branches with a more sophisticated information base for decision-making on employment matters, a speedy coordination of the manpower demand and supply in the region, and establishment of an inter-republican system that would provide the necessary information on vacancies and manpower demand.

4. Greater Social Securities Against Unemployment

It has been necessary to draft and approve the Fundamentals of Employment Legislation in view of substantial changes in the employment situation in the country after the Law on Property in the USSR was passed. The legal aspects of unemployment relief and reimbursements have not been sufficiently elaborated, since until recently the problem of unemployment did not exist. A number of resolutions were adopted at the outset of the reform that specified the rights of working people and the procedure for getting unemployment benefits and reimbursements. That was a big step forward in strengthening social security relating to the right to work. These are the most important legislative acts on the problems of employment passed during 1987 and 1988: the Resolution No. 1457 as of December 22, 1987 of the Central Committee of the CPSU, the USSR Council of Ministers and the All-Union Central Council of Trade Unions "On Ensuring Productive Employment, Improving the Employment System and Ensuring Social Guarantees of Working People"; the Decree of the Presidium of the Supreme Soviet of the RSFSR "Concerning Changes and Amendments to the Labor Code of the RSFSR" as of February 4, 1988; the regulations "On the Procedure of Dismissal and Employment of Labor and Office Workers and Payment of Benefits and Reimbursements", approved by the Secretariat of the All-Union Central Council of Trade Unions, as of March 2, 1988, the resolution of the State Committee of Labor of the USSR and the Secretariat of the All-Union Central Council of Trade Unions "On the Organization of Employment

Centers, Retraining, and Vocational Guidance of the Population" as of April 21, 1988.

The above documents specify the rights of redundant workers. They are entitled to another job at the same enterprise, or somewhere else according to the specialty, but if there is none, they are entitled to a different job according to their expressed wish and social needs, and there is also a possibility of new vocational training and corresponding employment afterwards. To expand employment possibilities, the most suitable ways of redistributing redundant workers are listed, and necessary and favorable conditions are provided for employment of women with children and with limited working ability (if they wish so) on part-time and home working bases, etc.

The criteria on which people may be dismissed from or kept in the work team have been specified. Priority access to jobs is given to workers who have a higher labor productivity and better skills irrespective of their age. The procedure of decision-making on staff cuts is to become more democratic and open, with the work team taking an active part in it.

It has been specified who is going to be responsible for employment and retraining of the redundant workers, and for paying out social benefits and reimbursements when people cannot be kept working at the enterprise. Special regulations have been adopted on the dismissal procedure, the responsibilities of the management, labor agencies and the superior administrative bodies which make the decision to close down or reorganize the enterprise, on providing jobs for the redundant workers in the same territory or in a different part of the country, or sending them if necessary for retraining, in which case their consent is necessary.

The role of the local Soviets of People's Deputies, employment agencies and trade-union councils in providing speedy employment for redundant workers, their vocational guidance and retraining has become greater. For instance, if a new employment cannot be provided during the time fixed by the legislation, the Soviets have the right to shift the date of the planned dismissal with the consent of a work team.

The procedure on vocation and career retraining of redundant workers has been consolidated. The enterprises which recruit redundant workers are to provide it either through their own training facilities or on a contractual basis, at vocational secondary and higher schools, other enterprises or organizations. The employment offices are to maintain data and provide enterprises with information about local vocational training courses and the number of people who intend to take a retraining course.

Proceeding from the development strategy of the industry, ministries and other administrative bodies have to foresee possible occupational redundancies and to plan measures to provide re-employment in the same industry or manpower redistribution to construction sites or new development zones.

A transition to a strong social policy which manifests itself in the working out and adaptation of redundancy benefits and reimbursements is a big step forward. The decision on bigger redundancy payments has been taken (it has risen from a fortnight's wages to the average monthly earnings). From now on a redundant worker while looking for a job is entitled to his average monthly wage during the second month from the day his employment terminated, or even during the third month, which is an exception made for persons who applied to the employment office within two weeks (to qualify for that exception a person must have two months' continuous unemployment since the date of application with no appropriate job offers in the given territory). If the enterprise is reorganized or closed down, redundant workers are entitled to their average monthly wages during the second and third months from the day of dismissal and above the fixed terminal wage.

A number of measures have been taken to preserve the workers' living standards during new career training and the period of adaptation to the new team. Those workers who are taking training courses during working hours are entitled to the average monthly wages, calculated on the basis of their former employment, or the difference from their average wages is reimbursed when a person takes a retraining course in non-working hours.

In actual fact the above mentioned provisions are not sufficient to solve the whole range of employment problems arising from a transition to a market economy. The thing is that not only should the interests of the redundant workers be protected but the right to work of other people who face difficulties getting a job must be also guaranteed. Redundancy payments from the enterprise itself, for a term of only three months, cannot be considered sufficient compensation for the loss of employment.

More radical measures aimed to secure the interests of people who lose their job or have difficulties in getting one are currently discussed. The essence of the discussion is that the state and the enterprise should share the responsibility to provide social security of employment, and a state system of retraining and employment should add to the social support of the redundant workers by enterprises, and a national employment fund should be established to ensure reimbursement to every unemployed able-bodied person who needs a job and income. New rules are needed to specify the mechanism that would set in action the national employment fund, the principles of community work, retraining, financing, relocation of people and their families to other parts of the country if they wish to get a job there and establishment of special employment zones in which the state will encourage the emergence of new jobs.

A discussion has been going on about such matters as the status of the unemployed, the terms of getting the unemployment benefits, the highest scales of relief, its time limits, material help to the dependents of the unemployed, and an all-round state control over the employment policy. The decision is still pending on the issue how to ensure equal rights for all the workers to social security in case of unemployment irrespective of the form of ownership of the means of the production in which they have been employed.

The transition to a real market is socially unacceptable till the reform that would ensure the right to work under the changing employment situation is finalized. The people are not psychologically prepared for such a transition, and setting it in motion is also impossible in view of the insufficient occupational mobility of the population. For these reasons, the moderately radical version of the economic reform with a corresponding regulation of the labor market seems more able to suit the interests of the people, since it coordinates the economic and social criteria of efficient employment.